Mathematics Practice Workbook Grade 8

Complete Content Review
Plus 2 Full-length Math Tests

Elise Baniam - Michael Smith

Mathematics Practice Workbook Grade 8

Published in the United State of America By

The Math Notion

Email: info@Mathnotion.com

Web: WWW.MathNotion.com

ISBN: 978-1-63620-114-6

About the Author

Elise Baniam has been a math instructor for over a decade now. She graduated in Mathematics. Since 2006, Elise has devoted his time to both teaching and developing exceptional math learning materials. As a Math instructor and test prep expert, Elise has worked with thousands of students. She has used the feedback of her students to develop a unique study program that can be used by students to drastically improve their math score fast and effectively.

– SAT Math Workbook

– ACT Math Workbook

– ISEE/SSAT Math Workbooks

– Common Core Math Workbooks

–many Math Education Workbooks

– and some Mathematics books …

As an experienced Math teacher, Mrs. Baniam employs a variety of formats to help students achieve their goals: she teaches students in large groups, and she provides training materials and textbooks through her website and through Amazon.

You can contact Elise via email at:

Elise@Mathnotion.com

Get the Targeted Practice You Need to Excel on the Mathematics Test Grade 8!

Mathematics Practice Grade 8 is **an excellent investment in your future** and the best solution for students who want to maximize their score and minimize study time. Practice is an essential part of preparing for a test and improving a test taker's chance of success. The best way to practice taking a test is by going through lots of math questions.

High-quality mathematics instruction ensures that students become problem solvers. We believe all students can develop deep conceptual understanding and procedural fluency in mathematics. In doing so, through this math workbook we help our students grapple with real problems, think mathematically, and create solutions.

Mathematics Practice Workbook allows you to:

- Reinforce your strengths and improve your weaknesses,

- Practice **2500+ realistic** math practice questions,

- Exercise math problems in a variety of formats that provide intensive practice,

- Review and study **Two Full-length Practice Tests** with detailed explanations,

...and much more!

This Comprehensive Math Practice Book is carefully designed to provide only that **clear and concise information** you need.

WWW.MathNotion.com

… So Much More Online!

✓ FREE Math Lessons

✓ More Math Learning Books!

✓ Mathematics Worksheets

✓ Online Math Tutors

For a PDF Version of This Book

Please Visit WWW.MathNotion.com

Contents

Chapter 1:

Whole Numbers

Add and Subtract Integers

Find the sum or difference.

1) $(+152) + (-98) =$

2) $(+74) + (-42) =$

3) $188 - 85 =$

4) $(-214) + 157 =$

5) $(-72) + 425 =$

6) $182 + (-265) =$

7) $(-15) + 38 =$

8) $415 - 310 =$

9) $(-18) - (-77) =$

10) $(-88) + (-57) =$

11) $(-124) - 304 =$

12) $1,420 - (-257) =$

13) $4 + (-15) + (-35) + (-15) =$

14) $(-17) + (-24) + 42 + 12 =$

15) $(-5) - 7 + 38 - 21 =$

16) $8 + (-19) + (-29 - 21) =$

17) $(+45) + (+28) + (-57) =$

18) $(-42) + (-32) =$

19) $-14 - 18 - 9 - 31 =$

20) $9 + (-28) =$

21) $134 - 90 - 53 - (-42) =$

22) $(+37) - (-9) =$

23) $(+7) - (+11) - (-19) =$

24) $(+27) - (+9) - (-32) =$

Multiplication and Division

Calculate.

1) $210 \times 9 =$

2) $160 \times 40 =$

3) $(-6) \times 8 \times (-5) =$

4) $-5 \times (-7) \times (-7) =$

5) $13 \times (-13) =$

6) $40 \times (-8) =$

7) $8 \times (-2) \times 6 =$

8) $(-400) \times (-30) =$

9) $(-20) \times (-20) \times 3 =$

10) $125 \times 6 =$

11) $142 \times 50 =$

12) $364 \div 14 =$

13) $(-4,125) \div 5 =$

14) $(-28) \div (-7) =$

15) $288 \div (-18) =$

16) $3,500 \div 28 =$

17) $(-126) \div 3 =$

18) $4,128 \div 4 =$

19) $1,260 \div (-35) =$

20) $3,360 \div 4 =$

21) $(-54) \div 2 =$

22) $(-2,000) \div (-20) =$

23) $0 \div 870 =$

24) $(-1,020) \div 6 =$

25) $5,868 \div 652 =$

26) $(-2,520) \div 4 =$

27) $10,902 \div 3 =$

28) $(-60) \div (-5) =$

Absolute Value

Simplify each equation below.

1) $|-40| =$

2) $-20 + |-40| + 38 =$

3) $|-58| - |-32| + 16 =$

4) $|-8 + 7 - 4| + |5 + 5| =$

5) $3|3 - 9| + 18 =$

6) $|-8| + |-25| =$

7) $|-42 + 18| + 12 - 5 =$

8) $|-16| - |-28| - 7 =$

9) $|-35| - |-14| + 9 =$

10) $|24| - 22 + |-15| =$

11) $\frac{3|4-8|}{4} =$

12) $|-26 + 11| =$

13) $|-24| \times |3| + 4 =$

14) $|-4| + |-28| + 6 - 2 =$

15) $|-16| + |-18| - 19 =$

16) $14 + |-28 + 12| + |-15| =$

17) $26 - |-56| + 20 =$

18) $\frac{|147|}{|7|} + 9 =$

19) $|-6 + 10| + |38 - 18| + 2 =$

20) $|-30 + 18| + |-15| + 10 =$

21) $\frac{|-54|}{6} \times |-9| =$

22) $\frac{6|3 \times 5|}{6} \times \frac{|-12|}{5} =$

23) $\frac{|3 \times 5|}{15} \times 8 =$

24) $|-20 + 4| \times \frac{|-2 \times 6|}{8} =$

25) $|-150 + 10| - 6 + 6 =$

26) $|-80 + 60| - 20 =$

Ordering Integers and Numbers

Order each set of integers from least to greatest.

1) $6, -9, -4, -3, 2$ ___, ___, ___, ___, ___, ___

2) $-4, -18, 5, 14, 11$ ___, ___, ___, ___, ___, ___

3) $29, -28, -16, 27, -21$ ___, ___, ___, ___, ___, ___

4) $-15, -45, 25, -17, 38$ ___, ___, ___, ___, ___, ___

5) $37, -42, 32, -45, 18$ ___, ___, ___, ___, ___, ___

6) $75, 38, -59, 85, -24$ ___, ___, ___, ___, ___, ___

Order each set of integers from greatest to least.

7) $16, 19, -11, -15, -7$ ___, ___, ___, ___, ___, ___

8) $22, 46, -54, -36, 61$ ___, ___, ___, ___, ___, ___

9) $55, -46, -19, 37, -17$ ___, ___, ___, ___, ___, ___

10) $37, 95, -46, -22, 87$ ___, ___, ___, ___, ___, ___

11) $-9, 79, -65, -78, 84$ ___, ___, ___, ___, ___, ___

12) $-70, -35, -50, 17, 39$ ___, ___, ___, ___, ___, ___

Order of Operations

Evaluate each expression.

1) $6 + (2 \times 5) =$

2) $15 - (4 \times 2) =$

3) $(14 \times 5) + 15 =$

4) $(18 - 3) - (4 \times 5) =$

5) $32 + (18 \div 3) =$

6) $(18 \times 4) \div 6 =$

7) $(63 \div 7) \times (-3) =$

8) $(7 \times 8) + (34 - 18) =$

9) $80 + (3 \times 3) + 5 =$

10) $(20 \times 8) \div (4 + 4) =$

11) $(-7) + (12 \times 5) + 11 =$

12) $(5 \times 9) - (45 \div 5) =$

13) $(7 \times 6 \div 2) - (17 + 13) =$

14) $(14 + 6 - 16) \times 8 - 12 =$

15) $(36 - 18 + 30) \times (96 \div 8) =$

16) $24 + \left(14 - (36 \div 6)\right) =$

17) $(7 + 10 - 4 - 9) + (24 \div 3) =$

18) $(90 - 15) + (16 - 18 + 8) =$

19) $(20 \times 3) + (16 \times 4) - 10 =$

20) $15 + 12 - (26 \times 5) + 15 =$

Factoring

Factor, write prime if prime.

1) 16

2) 82

3) 28

4) 46

5) 62

6) 65

7) 38

8) 10

9) 54

10) 85

11) 45

12) 90

13) 81

14) 55

15) 92

16) 114

17) 86

18) 70

19) 105

20) 80

21) 95

22) 36

23) 84

24) 110

25) 34

26) 98

27) 63

28) 106

Great Common Factor (GCF)

Find the GCF of the numbers.

1) 6, 15

2) 46, 27

3) 48, 58

4) 20, 25

5) 16, 36

6) 32, 42

7) 60, 25

8) 90, 35

9) 72, 9

10) 45, 54

11) 88, 54

12) 35, 70

13) 70, 20

14) 32, 82

15) 48, 96

16) 30, 85

17) 16, 24

18) 80, 100, 40

19) 81, 112

20) 56, 88

21) 10, 5, 25

22) 8, 18, 24

23) 15, 45, 60

24) 51, 33

Least Common Multiple (LCM)

Find the LCM of each.

1) 8, 10

2) 32, 16

3) 5, 10, 15

4) 14, 21

5) 20, 4, 30

6) 25, 5

7) 12, 60, 24

8) 5, 6

9) 13, 26, 54

10) 28, 35

11) 27, 54

12) 110, 22

13) 30, 15, 60

14) 18, 63

15) 40, 8, 5

16) 81, 18

17) 38, 19

18) 22, 44

19) 25, 60

20) 16, 48

21) 17, 10

22) 8, 28

23) 35, 70

24) 21, 6

Divisibility Rule

Apply the divisibility rules to find the factors of each number.

1) 15	2, 3, 4, 5, 6, 9, 10	13) 34	2, 3, 4, 5, 6, 9, 10
2) 124	2, 3, 4, 5, 6, 9, 10	14) 385	2, 3, 4, 5, 6, 9, 10
3) 352	2, 3, 4, 5, 6, 9, 10	15) 915	2, 3, 4, 5, 6, 9, 10
4) 94	2, 3, 4, 5, 6, 9, 10	16) 157	2, 3, 4, 5, 6, 9, 10
5) 241	2, 3, 4, 5, 6, 9, 10	17) 540	2, 3, 4, 5, 6, 9, 10
6) 455	2, 3, 4, 5, 6, 9, 10	18) 340	2, 3, 4, 5, 6, 9, 10
7) 65	2, 3, 4, 5, 6, 9, 10	19) 480	2, 3, 4, 5, 6, 9, 10
8) 320	2, 3, 4, 5, 6, 9, 10	20) 3,750	2, 3, 4, 5, 6, 9, 10
9) 1,134	2, 3, 4, 5, 6, 9, 10	21) 660	2, 3, 4, 5, 6, 9, 10
10) 68	2, 3, 4, 5, 6, 9, 10	22) 286	2, 3, 4, 5, 6, 9, 10
11) 754	2, 3, 4, 5, 6, 9, 10	23) 158	2, 3, 4, 5, 6, 9, 10
12) 148	2, 3, 4, 5, 6, 9, 10	24) 456	2, 3, 4, 5, 6, 9, 10

Answer key Chapter 1

Add and Subtract Integers

1) 54	9) 59	17) 16
2) 32	10) −145	18)
3) 103	11) −428	19)
4) −57	12) 1,677	20)
5) 353	13) −61	21)
6) −83	14) 13	22) 46
7) 23	15)	23) 15
8) 105	16) −61	24) 50

Multiplication and Division

1) 1,890	11) 7,100	21) −27
2) 6,400	12) 26	22) 100
3) 240	13) −825	23) 0
4) −245	14) 4	24) −170
5) −169	15) −16	25) 9
6) −320	16) 125	26) −630
7) −96	17) −42	27) 3,634
8) 12,000	18) 1,032	28) 12
9) 1,200	19) −36	
10) 750	20) 840	

Absolute Value

1) 40	8) −19	15) 15
2) 58	9) 30	16) 45
3) 42	10) 17	17) −10
4) 15	11) 3	18) 30
5) 36	12) 15	19) 26
6) 33	13) 76	20) 37
7) 31	14) 36	21) 81

22) 36   24) 24   26) 0

23) 8   25) 140

Ordering Integers and Numbers

1) -9, −4, −3, 2, 6   7) 19, 16, −7, −11, −15

2) −18, −4, 5, 11, 14   8) 61, 46, 22, −36, −54

3) −28, −21, −16, 27, 29   9) 55, 37, −17, −19, −46

4) −45, −17, −15, 25, 38   10) 95, 87, 37, −22, −46

5) −45, −42, 18, 32, 37   11) 84, 79, −9, −65, −78

6) −59, −24, 38, 75, 85   12) 39, 17, −35, −50, −70

Order of Operations

1) 16	6) 12	11) 64	16) 32
2) 7	7) −27	12) 36	17) 12
3) 85	8) 72	13) −9	18) 81
4) −5	9) 94	14) 20	19) 114
5) 38	10) 20	15) 576	20) −88

Factoring

1) 1, 2, 4, 8, 16   15) 1, 2, 4, 23, 46, 92

2) 1, 2, 41, 82   16) 1, 2, 3, 6, 19, 38, 57, 114

3) 1, 2, 4, 7, 14, 28   17) 1, 2, 43, 86

4) 1, 2, 23, 46   18) 1, 2, 5, 7, 10, 14, 35, 70

5) 1, 2, 31, 62   19) 1, 3, 5, 7, 15, 21, 35, 105

6) 1, 5, 13, 65   20) 1, 2, 4, 5, 8, 10, 16, 20, 40, 80

7) 1, 2, 19, 38   21) 1, 5, 19, 95

8) 1, 2, 5, 10   22) 1, 2, 3, 4, 6, 9, 12, 18, 36

9) 1, 2, 3, 6, 9, 18, 27, 54   23) 1, 2, 3, 4, 6, 7, 12, 14, 21, 28, 42, 84

10) 1, 5, 17, 85   24) 1, 2, 5, 10, 11, 22, 55, 110

11) 1, 3, 5, 9, 15, 45   25) 1, 2, 17, 34

12) 1, 2, 3, 5, 6, 9, 10, 15, 18, 30, 45, 90   26) 1, 2, 7, 14, 49, 98

13) 1, 3, 9, 27, 81   27) 1, 3, 7, 9, 21, 63

14) 1, 5, 11, 55   28) 1, 2, 53, 106

Great Common Factor (GCF)

1) 3	9) 9	17) 8
2) 1	10) 9	18) 20
3) 2	11) 2	19) 1
4) 5	12) 35	20) 8
5) 4	13) 10	21) 5
6) 2	14) 2	22) 2
7) 5	15) 48	23) 15
8) 5	16) 5	24)

Least Common Multiple (LCM)

1) 40	9) 702	17) 38
2) 32	10) 140	18) 44
3) 30	11) 54	19) 300
4) 42	12) 110	20) 48
5) 60	13) 60	21) 170
6) 5	14) 126	22) 56
7) 120	15) 40	23) 70
8) 30	16) 162	24) 42

Divisibility Rule

1) 15	2, <u>3</u>, 4, <u>5</u>, 6, 9, 10	13) 34	<u>2</u>, 3, 4, 5, 6, 9, 10
2) 124	<u>2</u>, 3, <u>4</u>, 5, 6, 9, 10	14) 385	2, 3, 4, <u>5</u>, 6, 9, 10
3) 352	<u>2</u>, 3, <u>4</u>, 5, 6, 9, 10	15) 915	2, <u>3</u>, 4, <u>5</u>, 6, 9, 10
4) 94	<u>2</u>, 3, 4, 5, 6, 9, 10	16) 157	2, 3, 4, 5, 6, 9, 10
5) 241	2, 3, 4, 5, 6, 9, 10	17) 540	<u>2</u>, <u>3</u>, <u>4</u>, <u>5</u>, <u>6</u>, <u>9</u>, <u>10</u>
6) 455	2, 3, 4, <u>5</u>, 6, 9, 10	18) 340	<u>2</u>, 3, <u>4</u>, <u>5</u>, 6, 9, <u>10</u>
7) 65	2, 3, 4, <u>5</u>, 6, 9, 10	19) 480	<u>2</u>, <u>3</u>, <u>4</u>, <u>5</u>, <u>6</u>, 9, <u>10</u>
8) 320	<u>2</u>, 3, <u>4</u>, <u>5</u>, 6, 9, 10	20) 3,750	<u>2</u>, <u>3</u>, 4, <u>5</u>, <u>6</u>, 9, <u>10</u>
9) 1,134	<u>2</u>, <u>3</u>, 4, 5, <u>6</u>, <u>9</u>, 10	21) 660	<u>2</u>, <u>3</u>, <u>4</u>, <u>5</u>, <u>6</u>, 9, <u>10</u>
10) 68	<u>2</u>, 3, <u>4</u>, 5, 6, 9, 10	22) 286	<u>2</u>, 3, 4, 5, 6, 9, 10
11) 754	<u>2</u>, 3, 4, 5, 6, 9, 10	23) 158	<u>2</u>, 3, 4, 5, 6, 9, 10
12) 148	<u>2</u>, 3, <u>4</u>, 5, 6, 9, 10	24) 456	<u>2</u>, <u>3</u>, <u>4</u>, 5, <u>6</u>, 9, 10

Chapter 2:

Fundamental

Adding Fractions – Unlike Denominator

Add the fractions and simplify the answers.

1) $\frac{1}{4} + \frac{2}{3} =$

2) $\frac{4}{5} + \frac{1}{2} =$

3) $\frac{1}{4} + \frac{5}{7} =$

4) $\frac{8}{11} + \frac{1}{2} =$

5) $\frac{7}{18} + \frac{1}{3} =$

6) $\frac{13}{54} + \frac{5}{18} =$

7) $\frac{5}{8} + \frac{1}{6} =$

8) $\frac{3}{10} + \frac{1}{4} =$

9) $\frac{5}{11} + \frac{2}{4} =$

10) $\frac{1}{9} + \frac{4}{7} =$

11) $\frac{5}{18} + \frac{3}{8} =$

12) $\frac{7}{32} + \frac{3}{4} =$

13) $\frac{9}{65} + \frac{2}{5} =$

14) $\frac{8}{63} + \frac{3}{7} =$

15) $\frac{11}{64} + \frac{1}{4} =$

16) $\frac{4}{15} + \frac{2}{5} =$

17) $\frac{4}{7} + \frac{3}{6} =$

18) $\frac{5}{72} + \frac{2}{9} =$

19) $\frac{2}{15} + \frac{1}{25} =$

20) $\frac{5}{12} + \frac{3}{8} =$

21) $\frac{7}{88} + \frac{1}{8} =$

22) $\frac{7}{12} + \frac{2}{5} =$

23) $\frac{3}{72} + \frac{1}{4} =$

24) $\frac{2}{27} + \frac{1}{18} =$

Subtracting Fractions – Unlike Denominator

Solve each problem.

1) $\frac{3}{4} - \frac{1}{5} =$

2) $\frac{2}{3} - \frac{1}{4} =$

3) $\frac{5}{6} - \frac{3}{7} =$

4) $\frac{5}{6} - \frac{7}{12} =$

5) $\frac{6}{7} - \frac{3}{14} =$

6) $\frac{7}{12} - \frac{7}{18} =$

7) $\frac{17}{20} - \frac{2}{5} =$

8) $\frac{2}{3} - \frac{1}{16} =$

9) $\frac{6}{7} - \frac{4}{9} =$

10) $\frac{3}{8} - \frac{5}{32} =$

11) $\frac{5}{7} - \frac{4}{35} =$

12) $\frac{5}{6} - \frac{7}{30} =$

13) $\frac{6}{7} - \frac{4}{21} =$

14) $\frac{5}{3} - \frac{8}{15} =$

15) $\frac{2}{11} - \frac{3}{22} =$

16) $\frac{5}{6} - \frac{4}{54} =$

17) $\frac{7}{24} - \frac{7}{32} =$

18) $\frac{3}{4} - \frac{3}{5} =$

19) $\frac{1}{2} - \frac{2}{9} =$

20) $\frac{2}{3} - \frac{6}{11} =$

Converting Mix Numbers

Convert the following mixed numbers into improper fractions.

1) $3\frac{5}{6} =$

2) $5\frac{11}{15} =$

3) $4\frac{1}{3} =$

4) $2\frac{4}{7} =$

5) $7\frac{1}{4} =$

6) $3\frac{19}{21} =$

7) $5\frac{9}{10} =$

8) $4\frac{7}{12} =$

9) $3\frac{10}{11} =$

10) $6\frac{2}{5} =$

11) $8\frac{2}{3} =$

12) $2\frac{11}{12} =$

13) $3\frac{5}{6} =$

14) $4\frac{8}{11} =$

15) $7\frac{1}{4} =$

16) $5\frac{6}{11} =$

17) $8\frac{1}{5} =$

18) $3\frac{7}{12} =$

19) $6\frac{1}{22} =$

20) $3\frac{2}{3} =$

21) $7\frac{4}{5} =$

22) $4\frac{7}{8} =$

23) $6\frac{5}{6} =$

24) $12\frac{9}{10} =$

Converting improper Fractions

Convert the following improper fractions into mixed numbers

1) $\frac{62}{14} =$

2) $\frac{98}{37} =$

3) $\frac{49}{17} =$

4) $\frac{57}{23} =$

5) $\frac{71}{16} =$

6) $\frac{137}{42} =$

7) $\frac{120}{33} =$

8) $\frac{26}{5} =$

9) $\frac{33}{19} =$

10) $\frac{13}{2} =$

11) $\frac{39}{4} =$

12) $\frac{210}{65} =$

13) $\frac{76}{64} =$

14) $\frac{18}{7} =$

15) $\frac{110}{13} =$

16) $\frac{49}{4} =$

17) $\frac{122}{9} =$

18) $\frac{61}{12} =$

19) $\frac{37}{6} =$

20) $\frac{28}{9} =$

21) $\frac{5}{4} =$

22) $\frac{79}{13} =$

23) $\frac{41}{8} =$

24) $\frac{64}{7} =$

Addition Mix Numbers

Add the following fractions.

1) $1 \frac{1}{5} + 4 \frac{2}{5} =$

2) $5 \frac{3}{7} + 3 \frac{4}{7} =$

3) $2 \frac{2}{8} + 3 \frac{1}{8} =$

4) $5 \frac{5}{8} + 3 \frac{1}{2} =$

5) $2 \frac{9}{14} + 3 \frac{3}{12} =$

6) $6 \frac{2}{5} + 3 \frac{1}{2} =$

7) $2 \frac{8}{27} + 2 \frac{2}{18} =$

8) $2 \frac{3}{4} + 3 \frac{1}{3} =$

9) $4 \frac{5}{6} + 1 \frac{1}{6} =$

10) $3 \frac{5}{7} + 1 \frac{3}{7} =$

11) $4 \frac{1}{2} + 2 \frac{2}{5} =$

12) $5 \frac{1}{4} + 2 \frac{5}{6} =$

13) $5 \frac{1}{3} + 2 \frac{2}{3} =$

14) $3 \frac{5}{6} + 3 \frac{2}{12} =$

15) $4 \frac{3}{5} + 4 \frac{1}{2} =$

16) $5 \frac{2}{3} + 1 \frac{4}{7} =$

17) $4 \frac{5}{6} + 6 \frac{1}{4} =$

18) $2 \frac{2}{5} + 3 \frac{3}{8} =$

19) $3 \frac{1}{6} + 2 \frac{4}{9} =$

20) $5 \frac{3}{5} + 3 \frac{2}{3} =$

21) $4 \frac{5}{8} + 1 \frac{1}{3} =$

22) $6 \frac{1}{9} + 4 \frac{4}{5} =$

23) $2 \frac{2}{7} + 3 \frac{4}{5} =$

24) $3 \frac{1}{2} + 1 \frac{5}{7} =$

Subtracting Mix Numbers

Subtract the following fractions.

1) $7\frac{1}{3} - 6\frac{1}{3} =$

2) $4\frac{5}{8} - 4\frac{2}{8} =$

3) $8\frac{5}{9} - 7\frac{1}{9} =$

4) $4\frac{1}{4} - 1\frac{1}{3} =$

5) $3\frac{1}{3} - 2\frac{1}{6} =$

6) $8\frac{1}{2} - 3\frac{2}{5} =$

7) $7\frac{5}{8} - 3\frac{3}{8} =$

8) $9\frac{9}{13} - 4\frac{6}{13} =$

9) $5\frac{7}{12} - 2\frac{5}{12} =$

10) $4\frac{4}{7} - 1\frac{3}{7} =$

11) $7\frac{1}{5} - 2\frac{1}{10} =$

12) $4\frac{5}{6} - 2\frac{1}{6} =$

13) $6\frac{2}{45} - 1\frac{1}{5} =$

14) $4\frac{1}{2} - 2\frac{1}{4} =$

15) $14\frac{4}{5} - 11\frac{2}{5} =$

16) $6\frac{2}{4} - 1\frac{1}{4} =$

17) $4\frac{1}{5} - 2\frac{3}{5} =$

18) $5\frac{1}{8} - 2\frac{1}{2} =$

19) $6\frac{2}{3} - 1\frac{1}{9} =$

20) $4\frac{3}{5} - 4\frac{1}{15} =$

21) $9\frac{9}{11} - 5\frac{1}{2} =$

22) $8\frac{4}{5} - 2\frac{3}{20} =$

23) $3\frac{2}{3} - 2\frac{1}{9} =$

24) $7\frac{9}{14} - 3\frac{3}{14} =$

Simplify Fractions

Reduce these fractions to lowest terms

1) $\frac{15}{10} =$

2) $\frac{20}{30} =$

3) $\frac{28}{35} =$

4) $\frac{21}{28} =$

5) $\frac{6}{18} =$

6) $\frac{27}{63} =$

7) $\frac{16}{28} =$

8) $\frac{48}{60} =$

9) $\frac{8}{72} =$

10) $\frac{30}{12} =$

11) $\frac{45}{60} =$

12) $\frac{30}{90} =$

13) $\frac{18}{30} =$

14) $\frac{5}{20} =$

15) $\frac{16}{56} =$

16) $\frac{56}{84} =$

17) $\frac{88}{33} =$

18) $\frac{36}{135} =$

19) $\frac{21}{56} =$

20) $\frac{64}{56} =$

21) $\frac{140}{280} =$

22) $\frac{30}{155} =$

23) $\frac{210}{42} =$

24) $\frac{130}{520} =$

Multiplying Fractions

Find the product.

1) $\frac{4}{5} \times \frac{2}{6} =$

2) $\frac{4}{22} \times \frac{5}{8} =$

3) $\frac{8}{30} \times \frac{12}{16} =$

4) $\frac{9}{14} \times \frac{21}{36} =$

5) $\frac{14}{15} \times \frac{5}{7} =$

6) $\frac{16}{19} \times \frac{3}{4} =$

7) $\frac{4}{9} \times \frac{9}{8} =$

8) $\frac{47}{85} \times 0 =$

9) $\frac{5}{8} \times \frac{16}{6} =$

10) $\frac{28}{15} \times \frac{5}{7} =$

11) $\frac{32}{24} \times \frac{12}{16} =$

12) $\frac{6}{42} \times \frac{7}{36} =$

13) $\frac{13}{8} \times \frac{12}{4} =$

14) $\frac{10}{9} \times \frac{6}{5} =$

15) $\frac{35}{56} \times \frac{8}{7} =$

16) $\frac{16}{18} \times 9 =$

17) $\frac{5}{22} \times \frac{44}{15} =$

18) $\frac{10}{18} \times \frac{9}{20} =$

19) $\frac{7}{11} \times \frac{8}{21} =$

20) $\frac{26}{24} \times \frac{8}{52} =$

21) $\frac{6}{17} \times \frac{1}{12} =$

22) $\frac{20}{9} \times \frac{6}{100} =$

23) $\frac{8}{14} \times \frac{7}{72} =$

24) $\frac{50}{100} \times \frac{300}{400} =$

Multiplying Mixed Number

Multiply. Reduce to lowest terms.

1) $2\frac{3}{5} \times 1\frac{3}{4} =$

2) $1\frac{5}{6} \times 1\frac{1}{3} =$

3) $2\frac{3}{5} \times 1\frac{1}{7} =$

4) $3\frac{1}{7} \times 2\frac{1}{2} =$

5) $4\frac{3}{4} \times 1\frac{1}{4} =$

6) $3\frac{1}{2} \times 1\frac{4}{5} =$

7) $3\frac{3}{4} \times 1\frac{1}{2} =$

8) $5\frac{2}{3} \times 3\frac{1}{3} =$

9) $3\frac{2}{3} \times 3\frac{1}{2} =$

10) $2\frac{1}{3} \times 3\frac{1}{2} =$

11) $4\frac{3}{4} \times 3\frac{2}{3} =$

12) $3\frac{2}{4} \times 3\frac{1}{6} =$

13) $2\frac{2}{5} \times 1\frac{1}{3} =$

14) $2\frac{1}{3} \times 1\frac{1}{6} =$

15) $2\frac{2}{3} \times 3\frac{1}{2} =$

16) $2\frac{1}{8} \times 2\frac{2}{5} =$

17) $2\frac{1}{4} \times 1\frac{2}{3} =$

18) $2\frac{3}{5} \times 1\frac{1}{4} =$

19) $2\frac{3}{5} \times 1\frac{5}{6} =$

20) $3\frac{3}{5} \times 2\frac{3}{4} =$

21) $3\frac{3}{4} \times 1\frac{1}{3} =$

22) $2\frac{5}{8} \times 3\frac{1}{4} =$

Dividing Fractions

Divide these fractions.

1) $1 \div \frac{1}{5} =$

2) $\frac{7}{13} \div 7 =$

3) $\frac{5}{14} \div \frac{2}{5} =$

4) $\frac{15}{60} \div \frac{3}{4} =$

5) $\frac{2}{17} \div \frac{4}{17} =$

6) $\frac{4}{16} \div \frac{18}{24} =$

7) $0 \div \frac{1}{9} =$

8) $\frac{12}{16} \div \frac{8}{9} =$

9) $\frac{8}{12} \div \frac{4}{18} =$

10) $\frac{9}{14} \div \frac{3}{7} =$

11) $\frac{8}{15} \div \frac{25}{16} =$

12) $\frac{35}{12} \div \frac{15}{6} =$

13) $\frac{9}{15} \div \frac{9}{5} =$

14) $\frac{8}{18} \div \frac{40}{6} =$

15) $\frac{45}{21} \div \frac{9}{21} =$

16) $\frac{7}{30} \div \frac{63}{5} =$

17) $\frac{36}{8} \div \frac{18}{24} =$

18) $9 \div \frac{1}{2} =$

19) $\frac{48}{35} \div \frac{8}{7} =$

20) $\frac{3}{36} \div \frac{9}{6} =$

21) $\frac{4}{7} \div \frac{12}{14} =$

22) $\frac{8}{40} \div \frac{10}{5} =$

Dividing Mixed Number

Divide the following mixed numbers. Cancel and simplify when possible.

1) $4\frac{1}{4} \div 4\frac{1}{3} =$

2) $2\frac{1}{6} \div 1\frac{2}{2} =$

3) $5\frac{1}{3} \div 3\frac{3}{4} =$

4) $3\frac{1}{6} \div 3\frac{1}{5} =$

5) $5\frac{1}{6} \div 1\frac{2}{3} =$

6) $3\frac{3}{5} \div 2\frac{2}{6} =$

7) $4\frac{3}{5} \div 2\frac{1}{3} =$

8) $2\frac{4}{9} \div 1\frac{1}{9} =$

9) $3\frac{5}{6} \div 3\frac{1}{2} =$

10) $9\frac{1}{9} \div 3\frac{2}{3} =$

11) $2\frac{2}{7} \div 4\frac{1}{7} =$

12) $4\frac{3}{8} \div 1\frac{3}{4} =$

13) $5\frac{1}{8} \div 1\frac{1}{12} =$

14) $6\frac{3}{8} \div 3\frac{1}{3} =$

15) $4\frac{2}{5} \div 1\frac{1}{5} =$

16) $2\frac{1}{2} \div 2\frac{2}{9} =$

17) $7\frac{1}{6} \div 5\frac{3}{8} =$

18) $5\frac{1}{2} \div 4\frac{1}{3} =$

19) $4\frac{5}{7} \div 1\frac{1}{3} =$

20) $3\frac{5}{6} \div 1\frac{1}{4} =$

21) $9\frac{1}{2} \div 7\frac{1}{3} =$

22) $3\frac{1}{8} \div 1\frac{1}{9} =$

23) $4\frac{1}{4} \div 3\frac{3}{4} =$

24) $3\frac{1}{6} \div 3\frac{1}{3} =$

Comparing Fractions

Compare the fractions, and write >, < or =

1) $\frac{14}{3}$ _____ $\frac{24}{15}$

2) $\frac{32}{3}$ _____ $\frac{2}{5}$

3) $\frac{4}{9}$ _____ $\frac{2}{4}$

4) $\frac{12}{4}$ _____ $\frac{13}{9}$

5) $\frac{1}{8}$ _____ $\frac{2}{3}$

6) $\frac{10}{6}$ _____ $\frac{16}{7}$

7) $\frac{12}{13}$ _____ $\frac{7}{9}$

8) $\frac{20}{14}$ _____ $\frac{25}{3}$

9) $4\frac{1}{12}$ _____ $6\frac{1}{3}$

10) $8\frac{1}{6}$ _____ $3\frac{1}{8}$

11) $3\frac{1}{2}$ _____ $3\frac{1}{5}$

12) $7\frac{5}{8}$ _____ $7\frac{2}{9}$

13) $3\frac{2}{8}$ _____ $5\frac{3}{5}$

14) $\frac{1}{15}$ _____ $\frac{3}{7}$

15) $\frac{31}{25}$ _____ $\frac{19}{83}$

16) $\frac{12}{100}$ _____ $\frac{6}{62}$

17) $15\frac{1}{4}$ _____ $15\frac{1}{9}$

18) $\frac{1}{5}$ _____ $\frac{1}{9}$

19) $\frac{1}{7}$ _____ $\frac{1}{13}$

20) $\frac{1}{18}$ _____ $\frac{8}{15}$

21) $\frac{7}{22}$ _____ $\frac{9}{76}$

22) $\frac{4}{5}$ _____ $\frac{2}{5}$

23) $2\frac{17}{14}$ _____ $3\frac{3}{14}$

24) $3\frac{25}{4}$ _____ $4\frac{5}{4}$

Answer key Chapter 2

Adding Fractions – Unlike Denominator

1) $\frac{11}{12}$

2) $\frac{13}{10}$

3) $\frac{27}{28}$

4) $\frac{27}{22}$

5) $\frac{13}{18}$

6) $\frac{14}{27}$

7) $\frac{19}{24}$

8) $\frac{11}{20}$

9) $\frac{21}{22}$

10) $\frac{43}{63}$

11) $\frac{47}{72}$

12) $\frac{31}{32}$

13) $\frac{7}{13}$

14) $\frac{5}{9}$

15) $\frac{27}{64}$

16) $\frac{2}{3}$

17) $\frac{15}{14}$

18) $\frac{7}{24}$

19) $\frac{13}{75}$

20) $\frac{19}{24}$

21) $\frac{9}{44}$

22) $\frac{59}{60}$

23) $\frac{7}{24}$

24) $\frac{7}{54}$

Subtracting Fractions – Unlike Denominator

1) $\frac{11}{20}$

2) $\frac{5}{12}$

3) $\frac{17}{42}$

4) $\frac{1}{4}$

5) $\frac{9}{14}$

6) $\frac{7}{36}$

7) $\frac{9}{20}$

8) $\frac{29}{48}$

9) $\frac{26}{63}$

10) $\frac{7}{32}$

11) $\frac{3}{5}$

12) $\frac{3}{5}$

13) $\frac{2}{3}$

14) $\frac{17}{15}$

15) $\frac{1}{22}$

16) $\frac{41}{54}$

17) $\frac{7}{96}$

18) $\frac{3}{20}$

19) $\frac{5}{18}$

20) $\frac{4}{33}$

Converting Mix Numbers

1) $\frac{23}{6}$

2) $\frac{86}{15}$

3) $\frac{13}{3}$

4) $\frac{18}{7}$

5) $\frac{29}{4}$

6) $\frac{82}{21}$

7) $\frac{59}{10}$

8) $\frac{55}{12}$

9) $\frac{43}{11}$

10) $\frac{32}{5}$

11) $\frac{26}{3}$

12) $\frac{35}{12}$

13) $\frac{23}{6}$

14) $\frac{52}{11}$

15) $\frac{29}{4}$

16) $\frac{61}{11}$

17) $\frac{41}{5}$

18) $\frac{43}{12}$

19) $\frac{133}{22}$ 21) $\frac{39}{5}$ 23) $\frac{41}{6}$

20) $\frac{11}{3}$ 22) $\frac{39}{8}$ 24) $\frac{129}{10}$

Converting improper Fractions

1) $4\frac{6}{14}$ 9) $1\frac{14}{19}$ 17) $13\frac{5}{9}$

2) $2\frac{24}{37}$ 10) $6\frac{1}{2}$ 18) $5\frac{1}{12}$

3) $2\frac{15}{17}$ 11) $9\frac{3}{4}$ 19) $6\frac{1}{6}$

4) $2\frac{11}{23}$ 12) $3\frac{15}{65}$ 20) $3\frac{1}{9}$

5) $4\frac{7}{16}$ 13) $1\frac{12}{64}$ 21) $1\frac{1}{4}$

6) $3\frac{11}{42}$ 14) $2\frac{4}{7}$ 22) $6\frac{1}{13}$

7) $3\frac{21}{33}$ 15) $8\frac{6}{13}$ 23) $5\frac{1}{8}$

8) $5\frac{1}{5}$ 16) $12\frac{1}{4}$ 24) $9\frac{1}{7}$

Adding Mix Numbers

1) $5\frac{3}{5}$ 9) 6 17) $11\frac{1}{12}$

2) 9 10) $5\frac{1}{7}$ 18) $5\frac{31}{40}$

3) $5\frac{3}{8}$ 11) $6\frac{9}{10}$ 19) $5\frac{11}{18}$

4) $9\frac{1}{8}$ 12) $8\frac{1}{12}$ 20) $9\frac{4}{15}$

5) $5\frac{25}{28}$ 13) 8 21) $5\frac{23}{24}$

6) $9\frac{9}{10}$ 14) 7 22) $10\frac{41}{45}$

7) $4\frac{11}{27}$ 15) $9\frac{1}{10}$ 23) $6\frac{3}{35}$

8) $6\frac{1}{12}$ 16) $7\frac{5}{21}$ 24) $5\frac{3}{14}$

Subtracting Mix Numbers

1) 1 5) $1\frac{1}{6}$ 8) $5\frac{3}{13}$

2) $\frac{3}{8}$ 6) $5\frac{1}{10}$ 9) $3\frac{1}{6}$

3) $1\frac{4}{9}$ 7) $4\frac{1}{4}$ 10) $3\frac{1}{7}$

4) $2\frac{11}{12}$

11) $5\frac{1}{10}$

12) $2\frac{2}{3}$

13) $4\frac{38}{45}$

14) $2\frac{1}{4}$

15) $3\frac{2}{5}$

16) $5\frac{1}{4}$

17) $1\frac{3}{5}$

18) $2\frac{5}{8}$

19) $5\frac{5}{9}$

20) $\frac{8}{15}$

21) $4\frac{7}{22}$

22) $6\frac{13}{20}$

23) $1\frac{5}{9}$

24) $4\frac{3}{7}$

Simplify Fractions

1) $\frac{3}{2}$

2) $\frac{2}{3}$

3) $\frac{4}{5}$

4) $\frac{3}{4}$

5) $\frac{1}{3}$

6) $\frac{3}{7}$

7) $\frac{4}{7}$

8) $\frac{4}{5}$

9) $\frac{1}{9}$

10) $\frac{5}{2}$

11) $\frac{3}{4}$

12) $\frac{1}{3}$

13) $\frac{3}{5}$

14) $\frac{1}{4}$

15) $\frac{2}{7}$

16) $\frac{2}{3}$

17) $\frac{8}{3}$

18) $\frac{4}{15}$

19) $\frac{3}{8}$

20) $\frac{8}{7}$

21) $\frac{1}{2}$

22) $\frac{6}{31}$

23) 5

24) $\frac{1}{4}$

Multiplying Fractions

1) $\frac{4}{15}$

2) $\frac{5}{44}$

3) $\frac{1}{5}$

4) $\frac{3}{8}$

5) $\frac{2}{3}$

6) $\frac{12}{19}$

7) $\frac{1}{2}$

8) 0

9) $\frac{5}{3}$

10) $\frac{4}{3}$

11) 1

12) $\frac{1}{36}$

13) $\frac{39}{8}$

14) $\frac{4}{3}$

15) $\frac{5}{7}$

16) 8

17) $\frac{2}{3}$

18) $\frac{1}{4}$

19) $\frac{8}{33}$

20) $\frac{1}{6}$

21) $\frac{1}{34}$

22) $\frac{2}{15}$

23) $\frac{1}{18}$

24) $\frac{3}{8}$

Multiplying Mixed Number

1) $4\frac{11}{20}$

2) $2\frac{4}{9}$

3) $2\frac{34}{35}$

4) $7\frac{6}{7}$

5) $5\frac{15}{16}$

6) $6\frac{3}{10}$

7) $5\frac{5}{8}$

8) $18\frac{8}{9}$

9) $12\frac{5}{6}$

10) $8\frac{1}{6}$

11) $17\frac{5}{12}$

12) $11\frac{1}{12}$

13) $3\frac{1}{5}$

14) $2\frac{13}{18}$

15) $9\frac{1}{3}$

16) $5\frac{1}{10}$

17) $3\frac{3}{4}$

18) $3\frac{1}{4}$

19) $4\frac{23}{30}$

20) $9\frac{9}{10}$

21) 5

22) $8\frac{17}{32}$

Dividing Fractions

1) 5

2) $\frac{1}{13}$

3) $\frac{25}{28}$

4) $\frac{1}{3}$

5) $\frac{1}{2}$

6) $\frac{1}{3}$

7) 0

8) $\frac{27}{32}$

9) 3

10) $\frac{3}{2}$

11) $\frac{128}{375}$

12) $\frac{7}{6}$

13) $\frac{1}{3}$

14) $\frac{1}{15}$

15) 5

16) $\frac{1}{54}$

17) 6

18) 18

19) $\frac{6}{5}$

20) $\frac{1}{18}$

21) $\frac{2}{3}$

22) $\frac{1}{10}$

Dividing Mixed Number

1) $\frac{51}{52}$

2) $\frac{1}{12}$

3) $1\frac{19}{45}$

4) $\frac{95}{96}$

5) $3\frac{1}{10}$

6) $1\frac{19}{35}$

7) $1\frac{34}{35}$

8) $\frac{2}{81}$

9) $1\frac{2}{21}$

10) $2\frac{16}{33}$

11) $\frac{16}{29}$

12) $2\frac{1}{2}$

13) $4\frac{19}{26}$

14) $1\frac{73}{80}$

15) $3\frac{2}{3}$

16) $1\frac{1}{8}$

17) $1\frac{1}{3}$

18) $1\frac{7}{26}$

19) $3\frac{15}{28}$

20) $3\frac{1}{15}$

21) $1\frac{13}{44}$

22) $2\frac{13}{16}$ 23) $1\frac{2}{15}$ 24) $\frac{19}{20}$

Comparing Fractions

1) >	7) >	13) <	19) >
2) >	8) <	14) <	20) <
3) <	9) <	15) >	21) >
4) >	10) >	16) <	22) >
5) <	11) >	17) >	23) =
6) <	12) >	18) >	24) >

Chapter 3:
Decimal

Round Decimals

Round each number to the correct place value

1) 0.8<u>3</u> =

2) 3.<u>0</u>2 =

3) 7.<u>7</u>11 =

4) 0.<u>4</u>78 =

5) <u>8</u>.824 =

6) 0.0<u>7</u>8 =

7) 8.<u>1</u>3 =

8) 84.8<u>4</u>0 =

9) 2.5<u>3</u>8 =

10) 12.<u>2</u>97 =

11) 2.<u>0</u>8 =

12) 5.<u>3</u>24 =

13) 2.<u>1</u>32 =

14) 8.0<u>7</u>32 =

15) 5<u>5</u>.78 =

16) 2<u>8</u>.24 =

17) 5<u>2</u>7.156 =

18) 624.<u>7</u>88 =

19) 17.4<u>8</u>1 =

20) 9<u>4</u>.86 =

21) 4.3<u>0</u>67 =

22) 57.<u>0</u>86 =

23) 224.<u>2</u>24 =

24) 0.1<u>3</u>44 =

25) 0.00<u>6</u>9 =

26) 9.0<u>3</u>86 =

27) 35.5<u>4</u>22 =

28) 11.0<u>9</u>31 =

Decimals Addition

Add the following.

1)
$\begin{array}{r} 32.12 \\ + \ 24.28 \\ \hline \end{array}$

8)
$\begin{array}{r} 56.25 \\ + \ 22.35 \\ \hline \end{array}$

2)
$\begin{array}{r} 0.88 \\ + \ 0.21 \\ \hline \end{array}$

9)
$\begin{array}{r} 46.21 \\ + \ 10.07 \\ \hline \end{array}$

3)
$\begin{array}{r} 15.36 \\ + \ 10.87 \\ \hline \end{array}$

10)
$\begin{array}{r} 8.96 \\ + \ 11.23 \\ \hline \end{array}$

4)
$\begin{array}{r} 75.165 \\ + \ 4.105 \\ \hline \end{array}$

11)
$\begin{array}{r} 15.214 \\ + \ 11.251 \\ \hline \end{array}$

5)
$\begin{array}{r} 8.650 \\ + \ 7.82 \\ \hline \end{array}$

12)
$\begin{array}{r} 72.36 \\ + \ 5.32 \\ \hline \end{array}$

6)
$\begin{array}{r} 5.324 \\ + \ 2.138 \\ \hline \end{array}$

13)
$\begin{array}{r} 32.05 \\ + \ 8.54 \\ \hline \end{array}$

7)
$\begin{array}{r} 81.21 \\ + \ 15.85 \\ \hline \end{array}$

14)
$\begin{array}{r} 137.21 \\ + \ 2.75 \\ \hline \end{array}$

Decimals Subtraction

Subtract the following

1) $\begin{array}{r} 9.35 \\ - \ 3.52 \\ \hline \end{array}$

2) $\begin{array}{r} 75.35 \\ - \ 62.37 \\ \hline \end{array}$

3) $\begin{array}{r} 0.68 \\ - \ 0.4 \\ \hline \end{array}$

4) $\begin{array}{r} 11.245 \\ - \ 8.6 \\ \hline \end{array}$

5) $\begin{array}{r} 0.652 \\ - \ 0.09 \\ \hline \end{array}$

6) $\begin{array}{r} 75.25 \\ - \ 28.88 \\ \hline \end{array}$

7) $\begin{array}{r} 112.66 \\ - \ 88.98 \\ \hline \end{array}$

8) $\begin{array}{r} 32.56 \\ - \ 12.45 \\ \hline \end{array}$

9) $\begin{array}{r} 68.35 \\ - \ 59.98 \\ \hline \end{array}$

10) $\begin{array}{r} 6.985 \\ - \ 0.223 \\ \hline \end{array}$

11) $\begin{array}{r} 55.69 \\ - \ 45.32 \\ \hline \end{array}$

12) $\begin{array}{r} 12.352 \\ - \ 2.325 \\ \hline \end{array}$

13) $\begin{array}{r} 19.231 \\ - \ 4.128 \\ \hline \end{array}$

14) $\begin{array}{r} 128.98 \\ - \ 7.92 \\ \hline \end{array}$

Decimals Multiplication

Solve.

1) $\begin{array}{r} 3.1 \\ \times\, 3.4 \\ \hline \end{array}$

2) $\begin{array}{r} 7.5 \\ \times\, 4.5 \\ \hline \end{array}$

3) $\begin{array}{r} 5.04 \\ \times\, 3.04 \\ \hline \end{array}$

4) $\begin{array}{r} 88.09 \\ \times\, 100 \\ \hline \end{array}$

5) $\begin{array}{r} 23.9 \\ \times\, 10 \\ \hline \end{array}$

6) $\begin{array}{r} 35.62 \\ \times\, 5.5 \\ \hline \end{array}$

7) $\begin{array}{r} 32.75 \\ \times\, 11.3 \\ \hline \end{array}$

8) $\begin{array}{r} 2.65 \\ \times\, 8.35 \\ \hline \end{array}$

9) $\begin{array}{r} 12.05 \\ \times\, 0.04 \\ \hline \end{array}$

10) $\begin{array}{r} 24.04 \\ \times\, 8.08 \\ \hline \end{array}$

11) $\begin{array}{r} 12.34 \\ \times\, 11.2 \\ \hline \end{array}$

12) $\begin{array}{r} 6.37 \\ \times\, 0.02 \\ \hline \end{array}$

13) $\begin{array}{r} 9.4 \\ \times\, 0.14 \\ \hline \end{array}$

14) $\begin{array}{r} 15.4 \\ \times\, 6.05 \\ \hline \end{array}$

Decimal Division

Dividing Decimals.

1) $8 \div 10,000 =$

2) $4 \div 100 =$

3) $3.4 \div 100 =$

4) $0.002 \div 10 =$

5) $8 \div 64 =$

6) $3 \div 81 =$

7) $5 \div 45 =$

8) $9 \div 180 =$

9) $7 \div 1,000 =$

10) $0.6 \div 0.63 =$

11) $0.9 \div 0.009 =$

12) $0.6 \div 0.12 =$

13) $0.6 \div 0.42 =$

14) $0.4 \div 0.04 =$

15) $3.08 \div 10 =$

16) $9.4 \div 10 =$

17) $6.75 \div 100 =$

18) $18.3 \div 3.3 =$

19) $64.4 \div 4 =$

20) $0.4 \div 0.004 =$

21) $7.05 \div 3.5 =$

22) $0.08 \div 0.40 =$

23) $0.9 \div 7.6 =$

24) $0.09 \div 54 =$

25) $5.24 \div 0.5 =$

26) $0.025 \div 125 =$

Comparing Decimals

Write the Correct Comparison Symbol (>, < or =)

1) 1.42 _____ 2.42

2) 0.5 _____ 0.425

3) 13.6 _____ 13.600

4) 7.07 _____ 7.70

5) 0.922 _____ 0.92

6) 0.856 _____ 0.956

7) 4.34 _____ 4.242

8) 5.0025 _____ 5.025

9) 24.087 _____ 24.078

10) 7.12 _____ 7.29

11) 4.44 _____ 4.444

12) 0.09 _____ 0.18

13) 1.302 _____ 1.32

14) 9.56 _____ 9.0569

15) 0.33 _____ 0.033

16) 21.04 _____ 21.040

17) 0.250 _____ 0.35

18) 44.92 _____ 45.01

19) 0.085 _____ 0.805

20) 36.5 _____ 29.8

21) 7.89 _____ 10.2

22) 0.024 _____ 0.0204

23) 5.042 _____ 0.5042

24) 7.5 _____ 0.758

25) 6.5 _____ 0.659

26) 3.24 _____ 3.2400

27) 8.34 _____ 0.834

28) 2.0809 _____ 2.0890

Convert Fraction to Decimal

Write each as a decimal.

1) $\dfrac{50}{100} =$

2) $\dfrac{46}{100} =$

3) $\dfrac{8}{50} =$

4) $\dfrac{8}{32} =$

5) $\dfrac{8}{72} =$

6) $\dfrac{56}{100} =$

7) $\dfrac{4}{50} =$

8) $\dfrac{31}{48} =$

9) $\dfrac{27}{300} =$

10) $\dfrac{15}{55} =$

11) $\dfrac{16}{32} =$

12) $\dfrac{6}{16} =$

13) $\dfrac{3}{10} =$

14) $\dfrac{18}{250} =$

15) $\dfrac{24}{80} =$

16) $\dfrac{30}{40} =$

17) $\dfrac{68}{100} =$

18) $\dfrac{7}{35} =$

19) $\dfrac{87}{100} =$

20) $\dfrac{1}{100} =$

21) $\dfrac{6}{36} =$

22) $\dfrac{2}{80} =$

Convert Decimal to Percent

Write each as a percent.

1) $0.187 =$

2) $0.19 =$

3) $2.6 =$

4) $0.017 =$

5) $0.009 =$

6) $0.786 =$

7) $0.245 =$

8) $0.57 =$

9) $0.002 =$

10) $0.205 =$

11) $0.324 =$

12) $84.9 =$

13) $3.015 =$

14) $0.7 =$

15) $2.35 =$

16) $0.0367 =$

17) $0.0043 =$

18) $0.960 =$

19) $6.68 =$

20) $0.484 =$

21) $8.957 =$

22) $0.879 =$

23) $2.7 =$

24) $0.9 =$

25) $3.6 =$

26) $26.8 =$

27) $1.01 =$

28) $0.006 =$

Convert Fraction to Percent

Write each as a percent.

1) $\frac{1}{4} =$

2) $\frac{3}{8} =$

3) $\frac{7}{14} =$

4) $\frac{15}{35} =$

5) $\frac{12}{28} =$

6) $\frac{17}{68} =$

7) $\frac{8}{11} =$

8) $\frac{14}{30} =$

9) $\frac{6}{50} =$

10) $\frac{12}{48} =$

11) $\frac{5}{34} =$

12) $\frac{27}{10} =$

13) $\frac{24}{80} =$

14) $\frac{16}{25} =$

15) $\frac{16}{58} =$

16) $\frac{2}{22} =$

17) $\frac{32}{88} =$

18) $\frac{21}{36} =$

19) $\frac{18}{92} =$

20) $\frac{6}{60} =$

21) $\frac{24}{600} =$

22) $\frac{720}{360} =$

Answer key Chapter 3

Round Decimals

1) 0.8	11) 2.1	21) 4.31
2) 3.0	12) 5.3	22) 57.1
3) 7.7	13) 2.1	23) 224.2
4) 0.5	14) 8.07	24) 0.13
5) 9.0	15) 56.0	25) 0.007
6) 0.08	16) 28.0	26) 9.04
7) 8.1	17) 530.0	27) 35.54
8) 84.84	18) 624.8	28) 11.09
9) 2.54	19) 17.48	
10) 12.3	20) 95.0	

Decimals Addition

1) 56.4	6) 7.462	11) 26.465
2) 1.09	7) 97.06	12) 77.68
3) 26.23	8) 78.6	13) 40.59
4) 79.27	9) 56.28	14) 139.96
5) 16.47	10) 20.19	

Decimals Subtraction

1) 5.83	6) 46.37	11) 10.37
2) 12.98	7) 23.68	12) 10.027
3) 0.28	8) 20.11	13) 15.103
4) 2.645	9) 8.37	14) 121.06
5) 0.562	10) 6.762	

Decimals Multiplication

1) 10.54	6) 195.91	11) 138.208
2) 33.75	7) 370.075	12) 0.1274
3) 15.3216	8) 22.1275	13) 1.316
4) 8,809	9) 0.482	14) 93.17
5) 239	10) 194.2432	

Decimal Division

1) 0.0008	2) 0.04	3) 0.034

4) 0.0002	12) 5	20) 100
5) 0.125	13) 1.4285…	21) 2.01428…
6) 0.037….	14) 10	22) 0.2
7) 0.111…	15) 0.308	23) 0.1184…
8) 0.05	16) 0.94	24) 0.0016
9) 0.007	17) 0.0675	25) 10.48
10) 0.952…	18) 5.5454…	26) 0.0002
11) 100	19) 16.1	

Comparing Decimals

1) <	11) <	21) <
2) >	12) <	22) >
3) =	13) <	23) >
4) <	14) >	24) >
5) >	15) >	25) >
6) <	16) =	26) =
7) >	17) <	27) >
8) <	18) <	28) <
9) >	19) <	
10) >	20) >	

Convert Fraction to Decimal

1) 0.5	9) 0.09	17) 0.68
2) 0.46	10) 0.27	18) 0.2
3) 0.16	11) 0.5	19) 0.87
4) 0.25	12) 0.375	20) 0.01
5) 0.11	13) 0.3	21) 0.166
6) 0.56	14) 0.072	22) 0.025
7) 0.08	15) 0.3	
8) 0.646	16) 0.75	

Convert Decimal to Percent

1) 18.7%	4) 1.7%	7) 24.5%
2) 19%	5) 0.9%	8) 57%
3) 260%	6) 78.6%	9) 0.2%

10) 20.5%

11) 32.4%

12) 8,490%

13) 301.5%

14) 70%

15) 235%

16) 3.67%

17) 0.43%

18) 96%

19) 668%

20) 48.4%

21) 895.7%

22) 87.9%

23) 270%

24) 90%

25) 360%

26) 2,680%

27) 101%

28) 0.6%

Convert Fraction to Percent

1) 25%

2) 37.5%

3) 50%

4) 42.86%

5) 29.31%

6) 25%

7) 72.72%

8) 46.66%

9) 12%

10) 25%

11) 14.7%

12) 2.7%

13) 30%

14) 64%

15) 27.58%

16) 9.09%

17) 36.36%

18) 58.33%

19) 19.56%

20) 10%

21) 4%

22) 200%

Chapter 4:

Equations and Inequality

Distributive and Simplifying Expressions

Simplify each expression.

1) $6x + 2 - 8 =$

2) $-(-4 - 5x) =$

3) $(-3x + 4)(-2) =$

4) $(-2x)(x + 3) =$

5) $-2x + x^2 + 4x^2 =$

6) $7y + 7x + 8y - 5x =$

7) $-3x + 3y + 14x - 9y =$

8) $-2x - 5 + 8x + \frac{16}{4} =$

9) $5 - 8(x - 2) =$

10) $-5 - 5x + 3x =$

11) $(x - 3y)2 + 4y =$

12) $2.5x^2 \times (-5x) =$

13) $-4 - 2x^2 + 6x^2 =$

14) $8 + 14x^2 + 4 =$

15) $4(-2x - 7) + 10 =$

16) $(-x)(-2 + 3x) - x(7 + x) =$

17) $-3(6 + 12) - 3x + 5x =$

18) $-4(5 - 12x - 3x) =$

19) $3(-2x - 6) =$

20) $9 + 7x - 9 =$

21) $x(-2x + 8) =$

22) $5xy + 4x - 3y + x + 2y =$

23) $3(-x - 7) + 9 =$

24) $(-3x - 4) + 7 =$

25) $3x + 4y - 5 + 1 =$

26) $(-2 + 3x) - 3x(1 + 2x) =$

27) $(-3)(-3x - 3y) =$

28) $4(-x - 2) + 5 =$

Factoring Expressions

Factor the common factor out of each expression.

1) $12x - 6 =$

2) $5x - 15 =$

3) $\frac{45}{15}x - 15 =$

4) $7b - 28 =$

5) $4a^2 - 24a =$

6) $2xy - 10y =$

7) $5x^2y + 15x =$

8) $a^2 - 8a + 7ab =$

9) $2a^2 + 2ab =$

10) $4x + 20 =$

11) $24x - 36xy =$

12) $8x - 6 =$

13) $\frac{1}{4}x - \frac{3}{4}y =$

14) $7xy - \frac{14}{3}x =$

15) $3ab + 9c =$

16) $\frac{1}{3}x - \frac{4}{3} =$

17) $10x - 15xy =$

18) $x^2 + 8x =$

19) $4x^2 - 12y =$

20) $4x^3 + 3xy + x^2 =$

21) $21x - 14 =$

22) $20b - 60c + 20d =$

23) $24ab - 8ac =$

24) $ax - ay - 3x + 3y =$

25) $3ax + 4a + 9x + 12 =$

26) $x^2 - 10x =$

27) $9x^3 - 18x^2 =$

28) $5x^2 - 70xy =$

Evaluate One Variable Expressions

Evaluate each using the values given.

1) $x + 4x, x = 3$

2) $5(-6 + 3x), x = 1$

3) $4x + 7x, x = -3$

4) $5(2 - x) + 5, x = 3$

5) $6x + 4x - 10, x = 2$

6) $5x + 11x + 12, x = -1$

7) $5x - 2x - 4, x = 5$

8) $\frac{3(5x+8)}{9}, x = 2$

9) $2x - 85, x = 32$

10) $\frac{x}{18}, x = 108$

11) $7(3 + 2x) - 33, x = 5$

12) $7(x + 3) - 23, x = 4$

13) $\frac{x+(-6)}{-3}, x = -6$

14) $8(6 - 3x) + 5, x = 2$

15) $-11 - \frac{x}{5} + 3x, x = 10$

16) $5x + 11x, x = 1$

17) $-12x + 3(5 + 3x), x = -7$

18) $x + 11x, x = 0.5$

19) $\frac{(2x-2)}{6}, x = 13$

20) $3(-1 - 2x), x = 5$

21) $5x - (5 - x), x = 3$

22) $\left(-\frac{15}{x}\right) + 2 + x, x = 5$

23) $-\frac{x \times 5}{x}, x = 5$

24) $2(-1 - 3x), x = 2$

25) $2x^2 + 7x, x = 1$

26) $2(3x + 1) - 4(x - 5), x = 3$

27) $-6x - 4, x = -5$

28) $7x + 2x, x = 3$

Evaluate Two Variable Expressions

Evaluate the expressions.

1) $x + 4y, \ x = 5, y = 2$

2) $(-2)(-3x - 2y), \ x = 1, y = 2$

3) $4x + 2y, \ x = 10, y = 5$

4) $\frac{x-4}{y+1}, \ x = 8, y = 3$

5) $\frac{a}{4} - 6b, a = 32, b = 4$

6) $3x - 4(y - 8), \ x = 5, y = 3$

7) $3x + 2y - 10, \ x = 2, y = 10$

8) $-3x + 10 + 8y - 5, \ x = 2, y = 1$

9) $yx \div 3, \ x = 9, y = 9$

10) $a - b \div 3, \ a = 3, b = 12$

11) $6(x - y), \ x = 7, y = 4$

12) $5x - 4y, \ x = 5, y = 8$

13) $\frac{10}{a} + 3b, \ a = 5, b = 4$

14) $2x^2 + 4xy, \ x = 3, y = 5$

15) $8 - \frac{xy}{10} + y, \ x = 6, y = 5$

16) $7(3x - y), \ x = 7, y = -9$

17) $5x^2 - 3y^2, \ x = -1, y = 2$

18) $3x + \frac{y}{4}, \ x = 6, y = 16$

19) $4(4x - 2y), \ x = 3, y = 5$

20) $4x(y - \frac{1}{2}), \ x = 5, y = 4$

21) $5(x^2 - 2y), \ x = 3, y = 2$

22) $5xy, \ x = 2, y = 8$

23) $\frac{1}{3}y^3\left(y - \frac{1}{4}x\right), \ x = -4, y = 3$

24) $-3(x - 5y) - 2x, \ x = 4, y = 2$

25) $-2x + \frac{1}{6}xy, \ x = 3, y = 6$

26) $x^2 + xy^2, \ x = 5, y = 7$

27) $x - 2y + 8, \ x = 9, y = 6$

28) $\frac{xy}{2x+y}, \ x = 5, y = 4$

Graphing Linear Equation

Sketch the graph of each line.

1) $y = 2x - 5$

2) $y = -2x + 3$

3) $x - y = 0$

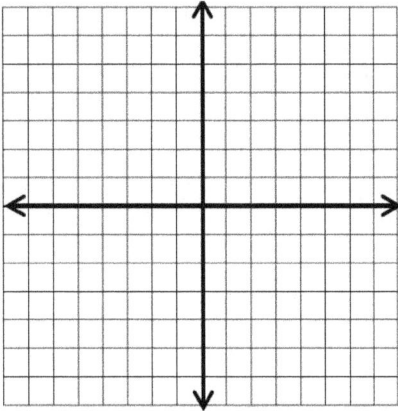 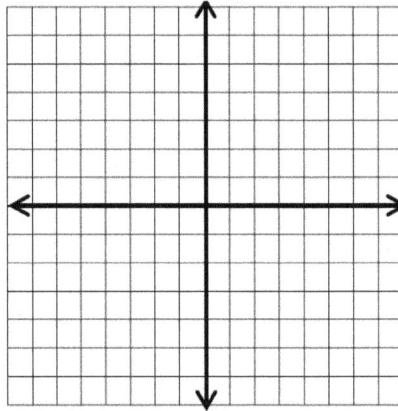 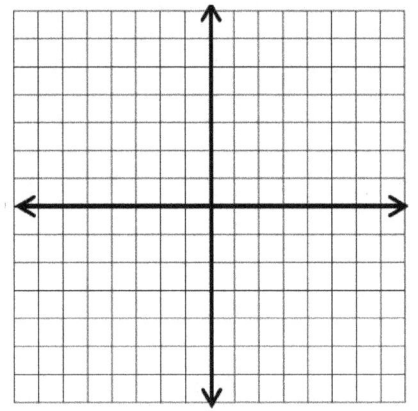

4) $x + y = 3$

5) $5x + 3y = -2$

6) $y - 3x + 2 = 0$

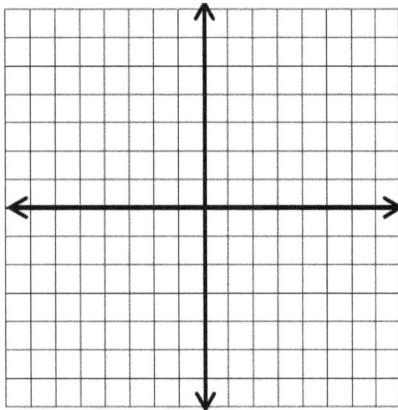 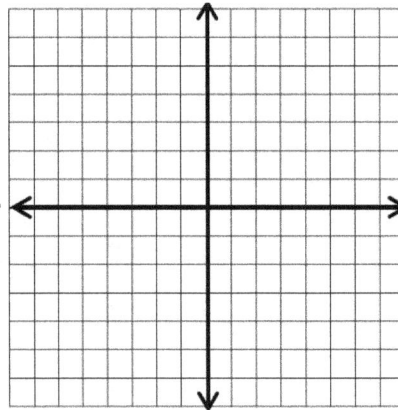 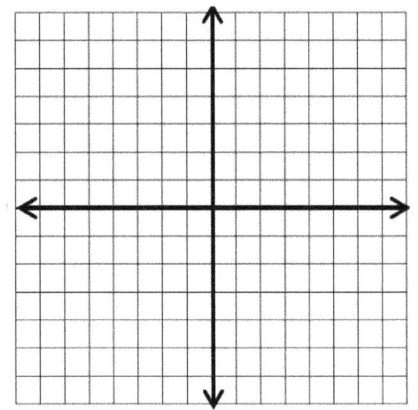

One Step Equations

Solve each equation.

1) $44 = (-12) + x$

2) $8x = (-64)$

3) $(-72) = (-8x)$

4) $(-5) = 3 + x$

5) $4 + \dfrac{x}{2} = (-3)$

6) $8x = (-104)$

7) $62 = x - 13$

8) $\dfrac{x}{3} = (-15)$

9) $x + 112 = 154$

10) $x - \dfrac{1}{3} = \dfrac{2}{3}$

11) $(-24) = x - 32$

12) $(-3x) = 39$

13) $(-169) = (13x)$

14) $-4x + 42 = 50$

15) $5x + 3 = 38$

16) $80 = (-8x)$

17) $3x + 7 = 19$

18) $24x = 144$

19) $x - 18 = 15$

20) $0.9x = 4.5$

21) $4x = 84$

22) $2x + 2.98 = 66.98$

23) $x + 9 = 6$

24) $x + 14 = 6$

25) $9x + 41 = 5$

26) $\dfrac{1}{4}x + 30 = 12$

Two Steps Equations

Solve each equation.

1) $6(3 + x) = 42$

2) $(-7)(x - 2) = 56$

3) $(-8)(3x - 4) = (-16)$

4) $5(2 + x) = -15$

5) $19(3x + 11) = 38$

6) $4(2x + 2) = 24$

7) $5(8 + 3x) = (-20)$

8) $(-5)(5x - 3) = 40$

9) $2x + 12 = 16$

10) $\frac{4x - 5}{5} = 3$

11) $(-3) = \frac{x + 4}{7}$

12) $80 = (-8)(x - 3)$

13) $\frac{x}{3} + 7 = 19$

14) $\frac{1}{4} = \frac{1}{2} + \frac{x}{4}$

15) $\frac{11 + x}{5} = (-6)$

16) $(-3)(10 + 5x) = (-15)$

17) $(-3x) + 12 = 24$

18) $\frac{x + 5}{5} = -5$

19) $\frac{x + 23}{8} = 3$

20) $(-4) + \frac{x}{2} = (-14)$

21) $-5 = \frac{x + 7}{8}$

22) $\frac{9x - 3}{6} = 4$

23) $\frac{2x - 12}{8} = 6$

24) $40 = (-5)(x - 8)$

Multi Steps Equations

Solve each equation.

1) $2 - (4 - 5x) = 3$

2) $-15 = -(4x + 7)$

3) $6x - 18 = (-2x) + 6$

4) $-32 = (-5x) - 11x$

5) $3(2 + 3x) + 3x = -30$

6) $5x - 18 = 2 + 2x - 7 + 2x$

7) $12 - 6x = (-36) - 3x + 3x$

8) $16 - 4x - 4x = 8 - 4x$

9) $8 + 7x + x = (-12) + 3x$

10) $(-3x) - 3(-2 + 4x) = 366$

11) $20 = (-200x) - 5 + 5$

12) $61 = 5x - 23 + 7x$

13) $7(4 + 2x) = 140$

14) $-60 = (-7x) - 13x$

15) $2(4x + 5) = -2(x + 4) - 22$

16) $11x - 17 = 6x + 8$

17) $9 = -3(x - 8)$

18) $(-6) - 8x = 6(1 + 2x)$

19) $x + 3 = -2(9 + 3x)$

20) $10 = 4 - 5x - 9$

21) $-15 - 9x - 3x = 12 - 3x$

22) $-23 - 3x + 5x = 27 - 23x$

23) $19 - 6x - 9x = -5 - 9x$

24) $15x - 18 = 6x + 9$

Graphing Linear Inequalities

Sketch the graph of each linear inequality.

1) $y > 2x - 3$

2) $y < x + 3$

3) $y \leq -3x - 8$

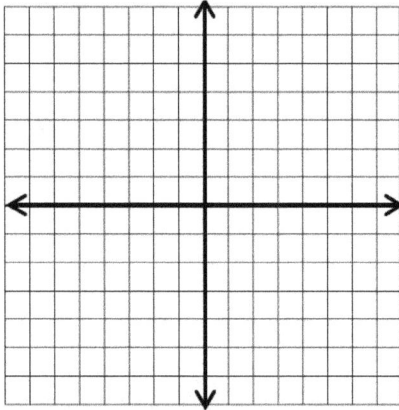 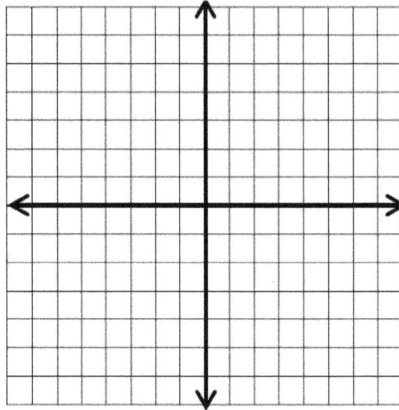 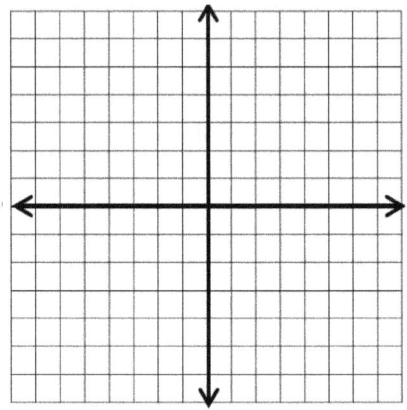

4) $3y \geq 6 + 3x$

5) $-3y < x - 12$

6) $2y \geq -8x + 4$

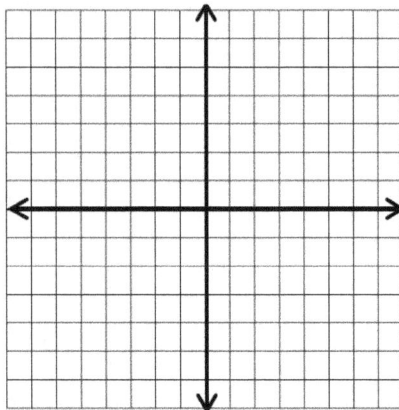 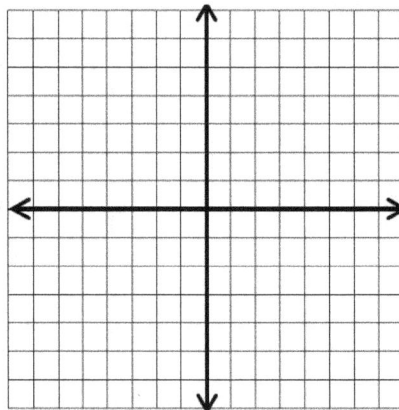 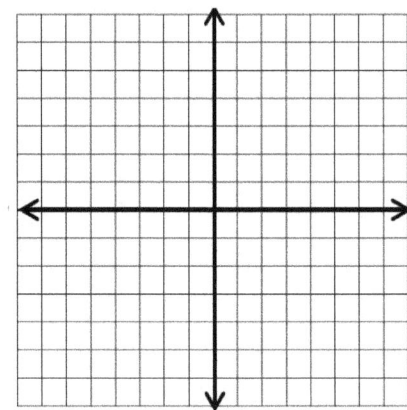

One Step Inequality

Solve each inequality.

1) $7x < 14$

2) $x + 7 \geq -8$

3) $x - 1 \leq 9$

4) $-2x + 4 > -10$

5) $x + 18 \geq -6$

6) $x + 9 \geq 5$

7) $x - \frac{1}{3} \leq 5$

8) $-7x < 42$

9) $-x + 8 > -3$

10) $\frac{x}{3} + 3 > -9$

11) $-x + 8 > -4$

12) $x - 14 \leq 18$

13) $-x - 5 \leq -7$

14) $x + 26 \geq -13$

15) $x + \frac{1}{3} \geq -\frac{2}{3}$

16) $x + 6 \geq -14$

17) $x - 42 \leq -48$

18) $x - 5 \leq 4$

19) $-x + 5 > -6$

20) $x + 6 \geq -12$

21) $8x + 6 \leq 22$

22) $4x - 3 \geq 9$

23) $3x - 5 < 22$

24) $6x - 8 \leq 40$

Two Steps Inequality

Solve each inequality

1) $2x - 3 \leq 7$

2) $3x - 4 \leq 8$

3) $\frac{-1}{4}x + \frac{x}{2} \leq \frac{1}{8}$

4) $5x + 10 \geq 30$

5) $4x - 7 \geq 9$

6) $3x - 5 \leq 16$

7) $8x - 2 \leq 14$

8) $9x + 5 \leq 23$

9) $2x + 10 > 32$

10) $\frac{x}{8} + 2 \leq 4$

11) $3x + 4 \geq 37$

12) $3x - 8 < 10$

13) $6 \geq \frac{x+7}{2}$

14) $3x + 9 < 48$

15) $\frac{4+x}{5} \geq 3$

16) $16 + 4x < 36$

17) $16 > 6x - 8$

18) $5 + \frac{x}{3} < 6$

19) $-4 + 4x > 24$

20) $5 + \frac{x}{7} < 3$

Multi Steps Inequality

Solve the inequalities.

1) $4x - 6 < 5x - 9$

2) $\frac{4x + 5}{3} \leq x$

3) $7x - 5 > 3x + 15$

4) $-3x > -6x + 4$

5) $3 + \frac{x}{2} < \frac{x}{4}$

6) $\frac{4x - 6}{8} > x$

7) $4x - 20 + 4 > 6x - 8$

8) $x - 8 > 11 + 3(x + 5)$

9) $\frac{x}{3} + 2 > x$

10) $-7x + 8 \geq -6(4x - 8) - 8x$

11) $7x - 4 \leq 8x + 9$

12) $\frac{2x - 7}{5} > 2$

13) $8(x + 2) < 6x + 10$

14) $-8x + 12 \leq 4(x - 9)$

15) $\frac{5x - 6}{3} > 3x + 2$

16) $2(x - 8) + 10 \geq 4x - 2$

17) $\frac{-5x + 7}{6} > 5x$

18) $-3x - 4 > -7x$

19) $\frac{1}{4}x - 12 > \frac{1}{8}x - 19$

20) $-4(x - 9) \leq 5x$

Systems of Equations

Calculate each system of equations.

1) $-6x + 7y = 8$ $x = \underline{\quad}$
    $x + 4y = 9$ $y = \underline{\quad}$

2) $-4x + 12y = 12$ $x = \underline{\quad}$
    $14x - 16y = 10$ $y = \underline{\quad}$

3) $y = -9$ $x = \underline{\quad}$
    $2x - 5y = 12$ $y = \underline{\quad}$

4) $4y = -4x + 20$ $x = \underline{\quad}$
    $8x - 2y = -12$ $y = \underline{\quad}$

5) $10x - 9y = -13$ $x = \underline{\quad}$
    $-5x + 3y = 11$ $y = \underline{\quad}$

6) $-6x - 8y = 10$ $x = \underline{\quad}$
    $4x - 8y = 20$ $y = \underline{\quad}$

7) $5x - 14y = -23$ $x = \underline{\quad}$
    $-6x + 7y = 8$ $y = \underline{\quad}$

8) $-4x + 3y = 3$ $x = \underline{\quad}$
    $-x + 2y = 5$ $y = \underline{\quad}$

9) $-4x + 5y = 15$ $x = \underline{\quad}$
    $-3x + 4y = -10$ $y = \underline{\quad}$

10) $-6x - 6y = -21$ $x = \underline{\quad}$
    $-6x + 6y = -66$ $y = \underline{\quad}$

11) $12x - 21y = 6$ $x = \underline{\quad}$
    $-6x - 3y = -12$ $y = \underline{\quad}$

12) $-4x - 4y = -14$ $x = \underline{\quad}$
    $4x - 4y = 44$ $y = \underline{\quad}$

13) $4x + 5y = 3$ $x = \underline{\quad}$
    $3x - y = 6$ $y = \underline{\quad}$

14) $3x - 2y = 2$ $x = \underline{\quad}$
    $10x - 10y = 20$ $y = \underline{\quad}$

15) $5x + 8y = 14$ $x = \underline{\quad}$
    $-3x - 2y = -3$ $y = \underline{\quad}$

16) $8x + 5y = 4$ $x = \underline{\quad}$
    $-3x - 4y = 15$ $y = \underline{\quad}$

Systems of Equations Word Problems

Find the answer for each word problem.

1) Tickets to a movie cost $6 for adults and $4 for students. A group of friends purchased 9 tickets for $50.00. How many adults ticket did they buy? ____

2) At a store, Eva bought two shirts and five hats for $77.00. Nicole bought three same shirts and four same hats for $84.00. What is the price of each shirt? _____

3) A farmhouse shelters 10 animals, some are pigs, and some are ducks. Altogether there are 36 legs. How many pigs are there? _____

4) A class of 85 students went on a field trip. They took 24 vehicles, some cars and some buses. If each car holds 3 students and each bus hold 16 students, how many buses did they take? _____

5) A theater is selling tickets for a performance. Mr. Smith purchased 8 senior tickets and 10 child tickets for $248 for his friends and family. Mr. Jackson purchased 4 senior tickets and 6 child tickets for $132. What is the price of a senior ticket? $_____

6) The difference of two numbers is 15. Their sum is 33. What is the bigger number? $_____

7) The sum of the digits of a certain two–digit number is 7. Reversing its digits increase the number by 9. What is the number? _____

8) The difference of two numbers is 11. Their sum is 25. What are the numbers? _____

9) The length of a rectangle is 5 meters greater than 2 times the width. The perimeter of rectangle is 28 meters. What is the length of the rectangle? _____

10) Jim has 23 nickels and dimes totaling $2.40. How many nickels does he have? _____

Finding Distance of Two Points

Find the distance between each pair of points.

1) $(2, 1), (-1, -3)$

2) $(-4, -2), (4, 4)$

3) $(-3, 0), (15, 24)$

4) $(-4, -1), (1, 11)$

5) $(3, -2), (-6, -14)$

6) $(-6, 0), (-2, 3)$

7) $(3, 2), (11, 17)$

8) $(-6, -10), (6, -1)$

9) $(5, 9), (-11, -3)$

10) $(6, -2), (2, -6)$

11) $(3, 0), (18, 36)$

12) $(8, 4), (3, -8)$

13) $(4, 2), (-5, -10)$

14) $(-8, 10), (4, 40)$

15) $(8, 4), (-10, -20)$

16) $(-8, -2), (16, 8)$

17) $(3, 5), (-5, -10)$

18) $(-10, 20), (35, 45)$

Find the midpoint of the line segment with the given endpoints.

1) $(-2, -2), (4, 2)$

2) $(10, 4), (-2, 4)$

3) $(12, -2), (4, 10)$

4) $(-6, -5), (2, 1)$

5) $(3, -2), (5, -2)$

6) $(-10, -4), (6, -2)$

7) $(4, 1), (-4, 9)$

8) $(-5, 6), (-5, 2)$

9) $(-8, 8), (4, -2)$

10) $(1, 7), (5, -1)$

11) $(-9, 5), (5, 3)$

12) $(7, 10), (-3, -6)$

13) $(-8, 14), (-8, 2)$

14) $(16, 7), (6, -3)$

15) $(5, 6), (-3, 4)$

16) $(-9, -1), (-5, 7)$

17) $(17, 9), (5, 11)$

18) $(-8, -11), (18, -1)$

Answer key Chapter 4

Distributive and Simplifying Expressions

1) $6x - 6$

2) $4 + 5x$

3) $6x - 8$

4) $-2x^2 - 6x$

5) $5x^2 - 2x$

6) $2x + 15y$

7) $11x - 6y$

8) $6x - 1$

9) $-8x + 21$

10) $-2x - 5$

11) $2x - 2y$

12) $-12.5x^3$

13) $4x^2 - 4$

14) $14x^2 + 12$

15) $-8x - 18$

16) $-4x^2 - 5x$

17) $2x - 54$

18) $60x - 20$

19) $-6x - 18$

20) $7x$

21) $-2x^2 + 8x$

22) $5x + y + 5xy$

23) $-3x - 12$

24) $-3x + 3$

25) $3x + 4y - 4$

26) $-6x^2 - 2$

27) $9x + 9y$

28) $-4x - 3$

Factoring Expressions

1) $3(4x - 2)$

2) $5(x - 3)$

3) $3(x - 5)$

4) $7(b - 4)$

5) $4a(a - 6)$

6) $2y(x - 5)$

7) $5x(xy + 3)$

8) $a(a - 8 + 7b)$

9) $2a(a + b)$

10) $4(x + 5)$

11) $12x(2 - 3y)$

12) $2(4x - 3)$

13) $\frac{1}{4}(x - 3y)$

14) $7x(y - \frac{2}{3})$

15) $3(ab + 3c)$

16) $\frac{1}{3}(x - 4)$

17) $5x(2 - 3y)$

18) $x(x + 8)$

19) $4(x^2 - 3y)$

20) $x(4x^2 + 3y + x)$

21) $7(3x - 2)$

22) $20(b - 3c + d)$

23) $8a(3b - c)$

24) $(x - y)(a - 3)$

25) $(3x + 4)(a + 3)$

26) $x(x - 10)$

27) $9x^2(x - 2)$

28) $5x(x - 14y)$

Evaluate One Variable Expressions

1) 15

2) -15

3) -33

4) 0

5) 10

6) -4

7) 11

8) 6

9) -21

10) 6

11) 58

12) 26

13) 4

14) 5

15) 17

16) 16

17) 36

18) 6

19) 4

20) -33

21) 13

22) 4

23) -5

24) -14

25) 9   26) 28   27) 26   28) 27

Evaluate Two Variable Expressions

1) 13   9) 27   17) −7   24) 10

2) 14   10) 3   18) 22   25) −3

3) 50   11) 18   19) 8   26) 270

4) 1   12) 20   20) 70   27) 5

5) −16   13) 17   21) 25   28) $\frac{10}{7}$

6) 35   14) 78   22) 80

7) 16   15) 10   23) 36

8) 7   16) 210

Graphing Lines Using Line Equation

1) $y = 2x - 5$

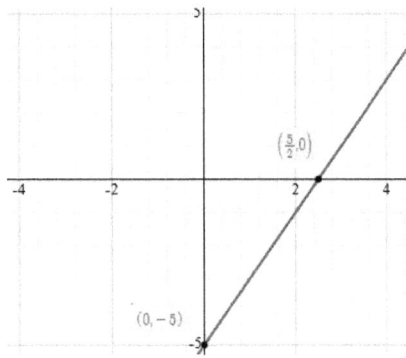

2) $y = -2x + 3$

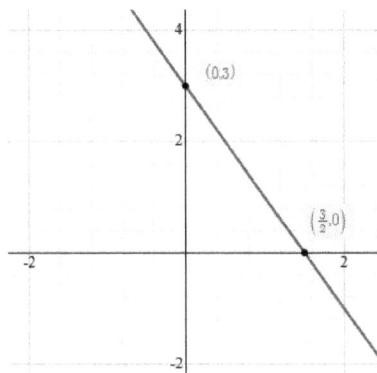

3) $x - y = 0$

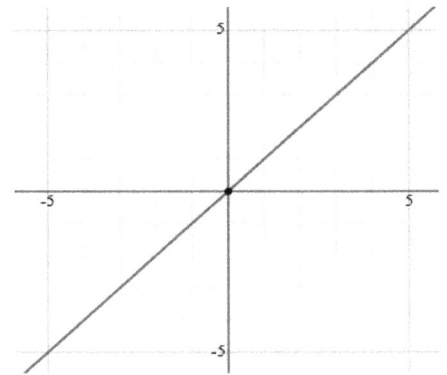

4) $x + y = 3$

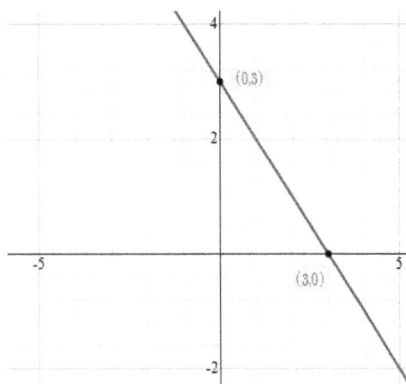

5) $5x + 3y = -2$

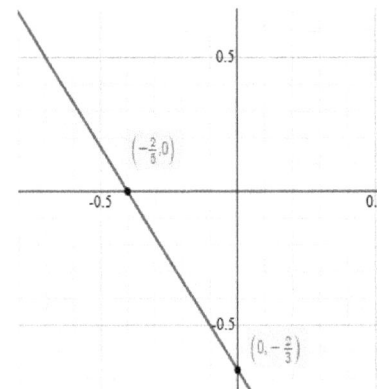

6) $y - 3x + 2 = 0$

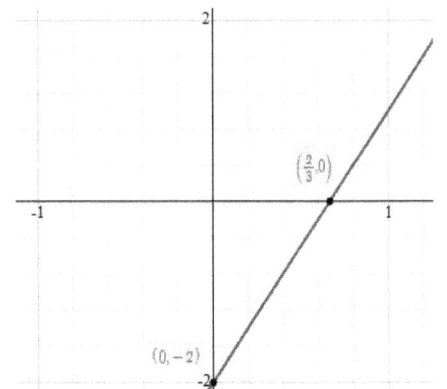

One Step Equations

1) $x = 56$

2) $x = -8$

3) $x = 9$

4) $x = -8$

5) $x = -14$

6) $x = -13$

7) x = 75

8) x = -45

9) x = 42

10) x = 1

11) x = 8

12) x = -13

13) x = -13

14) x = -2

15) x = 7

16) x = -10

17) $x = 4$

18) x = 6

19) x = 33

20) x = 5

21) x = 21

22) x = 32

23) $x = -3$

24)

25)

26)

Two Steps Equations

1) $x = 4$

2) $x = -6$

3) $x = 2$

4) $x = -5$

5) $x = -3$

6) $x = 2$

7) $x = -4$

8) $x = -1$

9) $x = 2$

10) $x = 5$

11) $x = -25$

12) $x = -7$

13) $x = 36$

14) $x = -1$

15) $x = -41$

16) $x = -1$

17) $x = -4$

18) $x = -30$

19) $x = 1$

20) $x = -20$

21) $x = -47$

22) $x = 3$

23) $x = 30$

24) $x = 0$

Multi Steps Equations

1) $x = 1$

2) $x = 2$

3) $x = 3$

4) $x = 2$

5) $x = -3$

6) $x = 13$

7) $x = 8$

8) $x = 2$

9) $x = -4$

10) $x = -24$

11) $x = -0.1$

12) $x = 7$

13) $x = 8$

14) $x = 3$

15) $x = -4$

16) $x = 5$

17) $x = 5$

18) $x = -3/5$

19) $x = -3$

20) $x = -3$

21) $x = -3$

22) $x = 2$

23) $x = 4$

24) $x = 3$

Graphing Linear Inequalities

1) $y > 2x - 3$

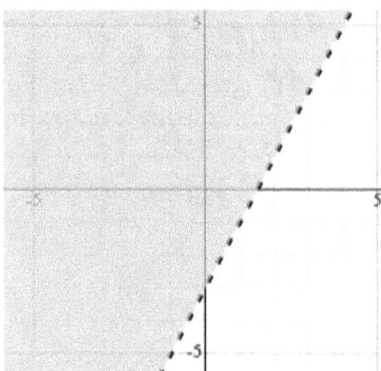

2) $y < x + 3$

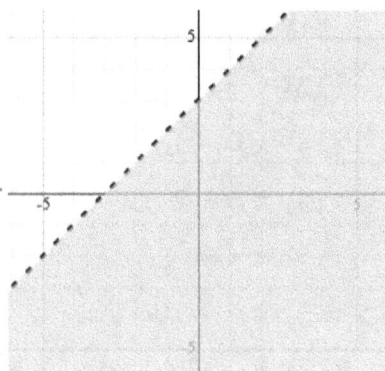

3) $y \le -3x - 8$

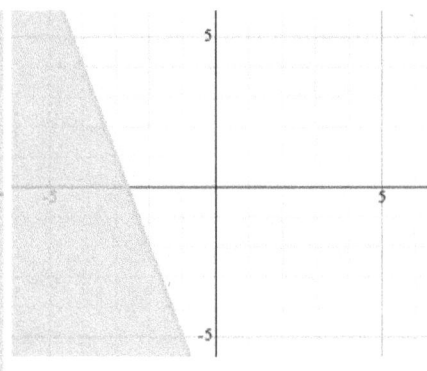

4) $3y \ge 6 + 3x$

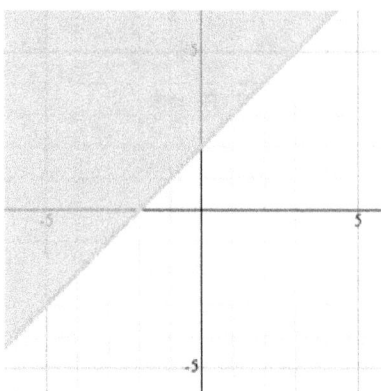

5) $-3y < x - 12$

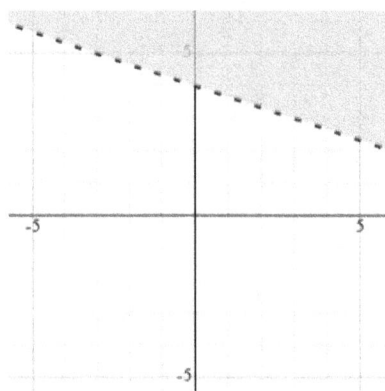

6) $2y \ge -8x + 4$

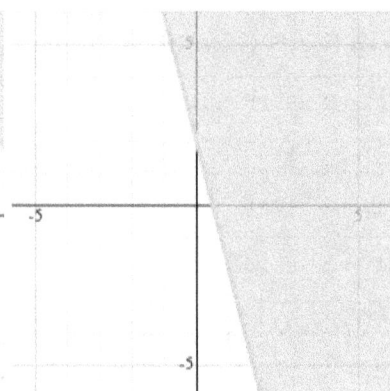

One Step Inequality

1) $x < 2$
2) $x \ge -15$
3) $x \le 10$
4) $x < 7$
5) $x \ge -24$
6) $x \ge -4$
7) $x \le \frac{16}{3}$
8) $x > -6$

9) $x < 11$
10) $x > -36$
11) $x < 12$
12) $x \le 32$
13) $x \ge 2$
14) $x \ge -39$
15) $x \ge -1$
16) $x \ge -20$

17) $x \le -6$
18) $x \le 9$
19) $x < 11$
20) $x \ge -18$
21) $x \le 2$
22) $x \ge 3$
23) $x < 9$
24) $x \le 8$

Two Steps Inequality

1) $x \le 5$
2) $x \le 4$
3) $x \le 0.5$

4) $x \ge 4$
5) $x \ge 4$
6) $x \le 7$

7) $x \le 2$
8) $x \le 2$
9) $x > 11$

10) $x \leq 16$

11) $x \geq 11$

12) $x < 6$

13) $x \leq 5$

14) $x < 13$

15) $x \geq 11$

16) $x < 5$

17) $x < 4$

18) $x < 3$

19) $x > 7$

20) $x < -14$

Multi Steps Inequality

1) $x > 3$

2) $x \leq -5$

3) $x > 5$

4) $x > \frac{4}{3}$

5) $x < -12$

6) $x < -1.5$

7) $x < -4$

8) $x < -17$

9) $x < 3$

10) $x \geq 1.6$

11) $x \geq -13$

12) $x > 8.5$

13) $x < -3$

14) $x \geq 4$

15) $x < -3$

16) $x \leq -2$

17) $x < \frac{1}{5}$

18) $x > 1$

19) $x > -56$

20) $x \geq 4$

Systems of Equations

1) $x = 1, y = 2$

2) $x = 3, y = -2$

3) $x = -\frac{33}{2}$

4) $x = -\frac{1}{5}, y = \frac{26}{5}$

5) $x = -4, y = -3$

6) $x = 1, y = -2$

7) $x = 1, y = 2$

8) $x = \frac{9}{5}, y = \frac{17}{5}$

9) $x = -110, y = -85$

10) $x = -\frac{15}{4}, y = \frac{29}{4}$

11) $x = \frac{5}{3}, y = \frac{2}{3}$

12) $x = -\frac{15}{4}, y = \frac{29}{4}$

13) $x = \frac{33}{19}, y = -\frac{15}{19}$

14) $x = -2, y = -4$

15) $x = -\frac{2}{7}, y = \frac{27}{14}$

16) $x = \frac{91}{17}, y = -\frac{132}{17}$

Systems of Equations Word Problems

1) 7

2) $16

3) 8

4) 1

5) $21

6) 24

7) 43

8) 18, 7

9) 11 meters

10) 18

Finding Distance of Two Points

1) 5

2) 10

3) 30

4) 13

5) 15

6) 5

7) 17

8) 15

9) 20

10) $4\sqrt{2}$

11) 39

12) 13

13) 15

14) $6\sqrt{29}$

15) 30

16) 26   17) 17   18) $5\sqrt{106}$

Finding Midpoint

1) $(1, 0)$ 7) $(0, 5)$ 13) $(-8, 8)$

2) $(4, 4)$ 8) $(-5, 4)$ 14) $(11, 2)$

3) $(8, 4)$ 9) $(-2, 3)$ 15) $(1, 5)$

4) $(-2, -2)$ 10) $(3, 3)$ 16) $(-7, 3)$

5) $(4, -2)$ 11) $(-2, 4)$ 17) $(11, 10)$

6) $(-2, -3)$ 12) $(2, 2)$ 18) $(5, -6)$

Chapter 5:

Exponent and Radicals

Positive Exponents

Simplify. Your answer should contain only positive exponents.

1) $3^4 =$

2) $2^5 =$

3) $\dfrac{3x^6y}{xy} =$

4) $(12x^2x)^3 =$

5) $(x^2)^4 =$

6) $\left(\dfrac{1}{4}\right)^3 =$

7) $0^8 =$

8) $6 \times 6 \times 6 =$

9) $3 \times 3 \times 3 \times 3 \times 3 =$

10) $(4x^4y)^2 =$

11) $9^3 =$

12) $(5x^3y^2)^2 =$

13) $5 \times 10^4 =$

14) $0.6 \times 0.6 \times 0.6 =$

15) $\dfrac{1}{3} \times \dfrac{1}{3} \times \dfrac{1}{3} =$

16) $4^5 =$

17) $(5x^8y^2)^3 =$

18) $7^3 =$

19) $y \times y \times y \times y =$

20) $8 \times 8 \times 8 \times 8 =$

21) $(2x^4y^2z)^3 =$

22) $8^0 =$

23) $(11x^4y^{-1})^4 =$

24) $(2x^2y^4)^5 =$

Negative Exponents

Simplify. Leave no negative exponents.

1) $2^{-4} =$

2) $9^{-2} =$

3) $\left(\frac{1}{3}\right)^{-2} =$

4) $8^{-3} =$

5) $1^{150} =$

6) $6^{-3} =$

7) $\left(\frac{1}{2}\right)^{-6} =$

8) $-8y^{-4} =$

9) $\left(\frac{1}{y^{-5}}\right)^{-3} =$

10) $x^{-\frac{4}{5}} =$

11) $\frac{1}{7^{-6}} =$

12) $3^{-5} =$

13) $5^{-2} =$

14) $13^{-2} =$

15) $30^{-2} =$

16) $x^{-8} =$

17) $(x^2)^{-4} =$

18) $x^{-2} \times x^{-2} \times x^{-2} \times x^{-2} =$

19) $\frac{1}{3} \times \frac{1}{3} =$

20) $100^{-2} =$

21) $100z^{-3} =$

22) $3^{-4} =$

23) $\left(-\frac{1}{11}\right)^{2} =$

24) $14^0 =$

25) $\left(\frac{1}{x}\right)^{-18} =$

26) $15^{-2} =$

Add and subtract Exponents

Solve each problem.

1) $4^2 + 5^3 =$

2) $x^8 + x^8 =$

3) $5b^3 - 4b^3 =$

4) $6 + 5^2 =$

5) $9 - 6^2 =$

6) $12 + 3^2 =$

7) $5x^2 + 8x^2 =$

8) $9^2 + 2^6 =$

9) $3^6 - 4^3 =$

10) $8^2 - 10^0 =$

11) $7^2 - 4^2 =$

12) $9^2 + 3^4 =$

13) $12^2 - 5^2 =$

14) $7^2 + 7^2 =$

15) $6^3 - 4^3 =$

16) $1^{24} + 1^{28} =$

17) $4^3 - 2^3 =$

18) $5^4 - 5^2 =$

19) $7^2 - 4^2 =$

20) $5^2 + 8^2 =$

21) $4^2 + 3^4 =$

22) $18 + 2^4 =$

23) $7x^8 + 5x^8 =$

24) $9^0 + 8^2 =$

25) $5^2 + 5^2 =$

26) $10^3 + 2^2 =$

27) $(\frac{1}{3})^2 + (\frac{1}{3})^2 =$

28) $8^2 + 2^2 =$

Exponent multiplication

Simplify each of the following

1) $2^5 \times 2^3 =$

2) $7^2 \times 8^0 =$

3) $9^1 \times 4^2 =$

4) $a^{-5} \times a^{-5} =$

5) $y^{-3} \times y^{-3} \times y^{-3} =$

6) $4^5 \times 5^7 \times 4^{-4} \times 5^{-6} =$

7) $6x^4y^3 \times 4x^3y^2 =$

8) $(x^3)^5 =$

9) $(x^4y^6)^5 \times (x^4y^5)^{-5} =$

10) $8^4 \times 8^2 =$

11) $a^{4b} \times a^0 =$

12) $4^2 \times 4^2 =$

13) $a^{3m} \times a^{2n} =$

14) $2a^n \times 4b^n =$

15) $5^{-3} \times 4^{-3} =$

16) $6^{10} \times 3^{10} =$

17) $(7^6)^5 =$

18) $(\frac{1}{6})^2 \times (\frac{1}{6})^4 \times (\frac{1}{6})^5 =$

19) $(\frac{1}{9})^{52} \times 9^{52} =$

20) $(4m)^{\frac{4}{5}} \times (-2m)^{\frac{4}{5}} =$

21) $(x^4y)^{\frac{1}{4}} \times (xy^3)^{\frac{1}{4}} =$

22) $(2a^m b^n)^r =$

23) $(5x^3y^2)^3 =$

24) $(x^{\frac{1}{3}}y^2)^{\frac{-1}{3}} \times (x^4y^6)^0 =$

25) $7^6 \times 7^5 =$

26) $28^{\frac{1}{6}} \times 28^{\frac{1}{3}} =$

27) $9^5 \times 3^5 =$

28) $(x^{12})^0 =$

Exponent division

Simplify. Your answer should contain only positive exponents.

1) $\dfrac{5^4}{5} =$

2) $\dfrac{38x^4}{x} =$

3) $\dfrac{a^m}{a^{2n}} =$

4) $\dfrac{3x^{-6}}{15x^{-4}} =$

5) $\dfrac{63x^9}{7x^4} =$

6) $\dfrac{17x^7}{5x^8} =$

7) $\dfrac{36x^8}{12y^3} =$

8) $\dfrac{45xy^6}{x^4y^2} =$

9) $\dfrac{3x^9}{8x} =$

10) $\dfrac{45x^7y^9}{5x^8} =$

11) $\dfrac{12x^4}{20x^9y^{12}} =$

12) $\dfrac{8yx^7}{40yx^{10}} =$

13) $\dfrac{21x^3y^2}{3x^2y^3} =$

14) $\dfrac{x^{4.75}}{x^{0.75}} =$

15) $\dfrac{9x^4y}{18xy^3} =$

16) $\dfrac{34b^3r^8}{17a^2b^5} =$

17) $\dfrac{30x^7}{15x^9} =$

18) $\dfrac{44x^5}{11x^8} =$

19) $\dfrac{6^5}{6^3} =$

20) $\dfrac{x}{x^{10}} =$

21) $\dfrac{13^7}{13^4} =$

22) $\dfrac{3xy^5}{12y^3} =$

23) $\dfrac{13x^6y}{169xy^3} =$

24) $\dfrac{48x^5}{8y^9} =$

Scientific Notation

Write each number in scientific notation.

1) 9,500,000=

2) 800 =

3) 0.000007 =

4) 387,000 =

5) 0.00139 =

6) 0.85 =

7) 0.000093 =

8) 20,000,000 =

9) 28,000,000 =

10) 230,000,000 =

11) 0.000049 =

12) 0.00002 =

13) 0.00027 =

14) 70,000 =

15) 2,870 =

16) 190,000 =

17) 0.0223 =

18) 0.7 =

19) 0.082 =

20) 310,000 =

21) 48,000 =

22) 0.000098 =

23) 0.035 =

24) 1,778 =

25) 58,781 =

26) 24,500 =

27) 33,021 =

28) 8,100,000 =

Square Roots

Find the square root of each number.

1) $\sqrt{64} =$

2) $\sqrt{0} =$

3) $\sqrt{324} =$

4) $\sqrt{484} =$

5) $\sqrt{1,600} =$

6) $\sqrt{529} =$

7) $\sqrt{0.01} =$

8) $\sqrt{10,000} =$

9) $\sqrt{0.16} =$

10) $\sqrt{0.36} =$

11) $\sqrt{0.25} =$

12) $\sqrt{1.21} =$

13) $\sqrt{784} =$

14) $\sqrt{576} =$

15) $\sqrt{676} =$

16) $\sqrt{961} =$

17) $\sqrt{1,681} =$

18) $\sqrt{0.81} =$

19) $\sqrt{0.49} =$

20) $\sqrt{0.64} =$

21) $\sqrt{1,089} =$

22) $\sqrt{2,500} =$

23) $\sqrt{8,100} =$

24) $\sqrt{12,100} =$

25) $\sqrt{2.25} =$

26) $\sqrt{1.69} =$

27) $\sqrt{1.44} =$

28) $\sqrt{0.04} =$

Simplify Square Roots

Simplify the following.

1) $\sqrt{54} =$

2) $\sqrt{108} =$

3) $\sqrt{12} =$

4) $\sqrt{99} =$

5) $\sqrt{200} =$

6) $\sqrt{45} =$

7) $8\sqrt{50} =$

8) $3\sqrt{300} =$

9) $\sqrt{24} =$

10) $2\sqrt{18} =$

11) $4\sqrt{3} + 7\sqrt{3} =$

12) $\frac{11}{4+\sqrt{5}} =$

13) $\sqrt{48} =$

14) $\frac{4}{3-\sqrt{5}} =$

15) $\sqrt{18} \times \sqrt{2} =$

16) $\frac{\sqrt{300}}{\sqrt{3}} =$

17) $\frac{\sqrt{90}}{\sqrt{18 \times 5}} =$

18) $\sqrt{80y^6} =$

19) $6\sqrt{81a} =$

20) $\sqrt{41+8} + \sqrt{9} =$

21) $\sqrt{72} =$

22) $\sqrt{432} =$

23) $\sqrt{112} =$

24) $\sqrt{128} =$

25) $\sqrt{768} =$

26) $\sqrt{96} =$

Answer key Chapter 5

Positive Exponents

1) 81
2) 32
3) $3x^5$
4) $1,728x^9$
5) x^8
6) $\frac{1}{64}$
7) 0
8) 6^3

9) 3^5
10) $16x^8y^2$
11) 729
12) $25x^6y^4$
13) 50,000
14) 0.6^3
15) $(\frac{1}{3})^3$
16) 1,024

17) $125x^{24}y^6$
18) 343
19) y^4
20) 8^4
21) $8x^{12}y^6z^3$
22) 1
23) $\frac{121x^8}{y^4}$
24) $32x^{10}y^{20}$

Negative Exponents

1) $\frac{1}{16}$
2) $\frac{1}{81}$
3) 9
4) $\frac{1}{512}$
5) 1
6) $\frac{1}{216}$
7) 64
8) $\frac{-8}{y^4}$
9) y^{15}
10) $\frac{1}{x^{\frac{4}{5}}}$

11) 7^6
12) $\frac{1}{243}$
13) $\frac{1}{25}$
14) $\frac{1}{169}$
15) $\frac{1}{900}$
16) $\frac{1}{x^8}$
17) $\frac{1}{x^8}$
18) $\frac{1}{x^8}$
19) $\frac{1}{3^2}$

20) $\frac{1}{10,000}$
21) $\frac{100}{z^3}$
22) $\frac{1}{81}$
23) $\frac{1}{121}$
24) 1
25) x^{18}
26) $\frac{1}{225}$

Add and subtract Exponents

1) 141
2) $2x^8$
3) b^3
4) 31
5) -27

6) 21
7) $13x^2$
8) 145
9) 665
10) 63

11) 33
12) 162
13) 119
14) 98
15) 152

16) 1

21) 97

26) 1,004

17) 56

22) 34

27) $\frac{2}{9}$

18) 600

23) $12x^8$

19) 33

24) 65

28) 68

20) 89

25) 50

Exponent multiplication

1) 2^8

11) a^{4b}

20) $(-8m^2)^{\frac{4}{5}}$

2) 49

12) 4^4

3) 144

13) a^{3m+2n}

21) $x^{\frac{5}{4}}y$

4) a^{-10}

14) $8(ab)^n$

22) $2^r a^{mr} b^{nr}$

5) y^{-9}

15) 20^{-3}

23) $125x^9y^6$

6) 20

16) 18^{10}

24) $x^{\frac{5}{4}}y$

7) $24x^7y^5$

17) 7^{30}

25) 7^{11}

8) x^{15}

18) $(\frac{1}{6})^{11}$

26) $28^{\frac{1}{2}}$

9) y^5

19) 1

27) $27^5 = 3^{15}$

10) 8^6

28) 1

Exponent division

1) 5^3

9) $\frac{3x^8}{8}$

17) $\frac{2}{x^2}$

2) $38x^3$

10) $\frac{9y^9}{x}$

18) $\frac{4}{x^3}$

3) a^{m-2n}

11) $\frac{3}{5x^5y^{12}}$

19) 6^2

4) $\frac{1}{5x^2}$

12) $\frac{1}{5x^3}$

20) $\frac{1}{x^9}$

5) $9x^5$

13) $\frac{7x}{y}$

21) 13^3

6) $\frac{17}{5x}$

14) x^4

22) $\frac{1}{4}xy^2$

7) $\frac{3x^8}{y^3}$

15) $\frac{x^3}{2y^2}$

23) $\frac{x^5}{13y^2}$

8) $\frac{45y^4}{x^3}$

16) $\frac{2r^8}{a^2b^2}$

24) $\frac{6x^5}{y^9}$

Scientific Notation

1) 9.5×10^6

3) 7×10^{-6}

5) 1.39×10^{-3}

2) 8×10^2

4) 3.87×10^5

6) 8.5×10^{-1}

7) 9.3×10^{-5}

8) 2×10^7

9) 2.8×10^7

10) 2.3×10^8

11) 4.9×10^{-5}

12) 2×10^{-5}

13) 2.7×10^{-4}

14) 7×10^4

15) 2.87×10^3

16) 1.9×10^5

17) 2.23×10^{-2}

18) 7×10^{-1}

19) 8.2×10^{-2}

20) 3.1×10^5

21) 4.8×10^4

22) 9.8×10^{-5}

23) 3.5×10^{-2}

24) 1.778×10^3

25) 5.8781×10^4

26) 2.45×10^4

27) 3.3021×10^4

28) 8.1×10^6

Square Roots

1) 8

2) 0

3) 18

4) 22

5) 40

6) 23

7) 0.1

8) 100

9) 0.4

10) 0.6

11) 0.5

12) 1.1

13) 28

14) 24

15) 26

16) 31

17) 41

18) 0.9

19) 0.7

20) 0.8

21) 33

22) 50

23) 90

24) 110

25) 1.5

26) 1.3

27) 1.2

28) 0.2

Simplify Square Roots

1) $3\sqrt{6}$

2) $6\sqrt{3}$

3) $2\sqrt{3}$

4) $3\sqrt{11}$

5) $10\sqrt{2}$

6) $3\sqrt{5}$

7) $40\sqrt{2}$

8) $30\sqrt{3}$

9) $2\sqrt{6}$

10) $6\sqrt{2}$

11) $11\sqrt{3}$

12) $4 - \sqrt{5}$

13) $4\sqrt{3}$

14) $3 + \sqrt{5}$

15) 6

16) 10

17) 1

18) $4y^3\sqrt{5}$

19) $54\sqrt{a}$

20) 10

21) $6\sqrt{2}$

22) $12\sqrt{3}$

23) $4\sqrt{7}$

24) $8\sqrt{2}$

25) $16\sqrt{3}$

26) $4\sqrt{6}$

Chapter 6:
Ratio, Proportion and Percent

Proportions

Find a missing number in a proportion.

1) $\dfrac{4}{7} = \dfrac{12}{a}$

11) $\dfrac{10}{6} = \dfrac{5}{a}$

2) $\dfrac{a}{9} = \dfrac{20}{45}$

12) $\dfrac{16}{a} = \dfrac{4}{19}$

3) $\dfrac{12}{60} = \dfrac{a}{5}$

13) $\dfrac{4}{11} = \dfrac{a}{12}$

4) $\dfrac{16}{a} = \dfrac{96}{36}$

14) $\dfrac{\sqrt{36}}{3} = \dfrac{48}{a}$

5) $\dfrac{4}{a} = \dfrac{16}{75}$

15) $\dfrac{6}{a} = \dfrac{6.6}{39.6}$

6) $\dfrac{\sqrt{9}}{4} = \dfrac{a}{32}$

16) $\dfrac{60}{140} = \dfrac{a}{280}$

7) $\dfrac{2}{4} = \dfrac{18}{a}$

17) $\dfrac{42}{200} = \dfrac{a}{68}$

8) $\dfrac{7}{14} = \dfrac{a}{35}$

18) $\dfrac{26}{104} = \dfrac{a}{4}$

9) $\dfrac{7}{a} = \dfrac{4.2}{6}$

19) $\dfrac{10}{16} = \dfrac{2}{a}$

10) $\dfrac{2}{12} = \dfrac{8}{a}$

20) $\dfrac{9}{7} = \dfrac{27}{a}$

Reduce Ratio

Reduce each ratio to the simplest form.

1) 5: 20 =

2) 6: 36 =

3) 63: 35 =

4) 24: 20 =

5) 12: 120 =

6) 16: 2 =

7) 70: 350 =

8) 4: 144 =

9) 25: 75 =

10) 4.8: 5.6 =

11) 110: 330 =

12) 3: 5 =

13) 120: 200 =

14) 30: 45 =

15) 34: 68 =

16) 32: 8 =

17) 140: 35 =

18) 20: 200 =

19) 126: 84 =

20) 156: 198 =

21) 40: 80 =

22) 42: 49 =

23) 5: 75 =

24) 18: 108 =

Percent

Find the Percent of Numbers.

1) 30% of 42 =

2) 28% of 15 =

3) 15% of 14 =

4) 24% of 70 =

5) 8% of 80 =

6) 35% of 12 =

7) 18% of 5 =

8) 12% of 46 =

9) 40% of 62 =

10) 4.5% of 50 =

11) 85% of 18 =

12) 60% of 50 =

13) 18% of 180 =

14) 2% of 240 =

15) 75% of 0 =

16) 80% of 120 =

17) 36% of 45 =

18) 10% of 70 =

19) 8% of 13 =

20) 4% of 8 =

21) 30% of 44 =

22) 80% of 17 =

23) 22% of 35 =

24) 8% of 150 =

25) 40% of 270 =

26) 2% of 5 =

27) 9% of 320 =

28) 10% of 26 =

Discount, Tax and Tip

Find the selling price of each item.

1) Original price of a computer: $150

 Tax: 8%, Selling price: $_____

2) Original price of a laptop: $240

 Tax: 15%, Selling price: $_____

3) Original price of a sofa: $300

 Tax: 8%, Selling price: $_____

4) Original price of a car: $12,600

 Tax: 3.5%, Selling price: $_____

5) Original price of a Table: $500

 Tax: 12%, Selling price: $_____

6) Original price of a house: $280,000

 Tax: 1.5%, Selling price: $_____

7) Original price of a tablet: $460

Discount: 30%, Selling price: $_____

8) Original price of a chair: $110

 Discount: 10%, Selling price: $_____

9) Original price of a book: $30

 Discount: 10% Selling price: $_____

10) Original price of a cellphone: 720

 Discount: 12% Selling price: $_____

11) Food bill: $27

 Tip: 15%   Price: $_____

12) Food bill: $45

 Tipp: 10%   Price: $_____

13) Food bill: $50

 Tip: 25%   Price: $_____

14) Food bill: $72

 Tipp: 18%   Price: $_____

Find the answer for each word problem.

15) Nicolas hired a moving company. The company charged $400 for its services, and Nicolas gives the movers a 25% tip. How much does Nicolas tip the movers? $_____

16) Mason has lunch at a restaurant and the cost of his meal is $80. Mason wants to leave a 8% tip. What is Mason's total bill including tip? $_____

Percent of Change

Find each percent of change.

1) From 300 to 600. ___ %

2) From 45 ft to 225 ft. ___ %

3) From $60 to $420. ___ %

4) From 50 cm to 150 cm. ___%

5) From 10 to 30. ___ %

6) From 60 to 108. ___ %

7) From 120 to 180. ___ %

8) From 400 to 600. ___ %

9) From 85 to 119. ___ %

10) From 100 to 175. ___ %

Calculate each percent of change word problem.

11) Bob got a raise, and his hourly wage increased from $32 to $40. What is the percent increase? ____ %

12) The price of a pair of shoes increases from $50 to $80. What is the percent increase? ___ %

13) At a coffee shop, the price of a cup of coffee increased from $3.50 to $4.2. What is the percent increase in the cost of the coffee? _____ %

14) 22cm are cut from a 88 cm board. What is the percent decrease in length? _ %

15) In a class, the number of students has been increased from 112 to 168. What is the percent increase? _____ %

16) The price of gasoline rose from $18.4 to $21.16 in one month. By what percent did the gas price rise? _____ %

17) A shirt was originally priced at $42. It went on sale for $33.6. What was the percent that the shirt was discounted? _____ %

Simple Interest

Determine the simple interest for these loans.

1) $140 at 18% for 5 years. $ _____

2) $1,800 at 6% for 2 years. $ _____

3) $900 at 25% for 4 years. $ _____

4) $9,200 at 1.5% for 8 months. $ ___

5) $600 at 5% for 7 months. $ _____

6) $40,000 at 8.5% for 3 years. $ ____

7) $7,400 at 8% for 5 years. $ _____

8) $500 at 2.5% for 2 years. $ _____

9) $600 at 6.5 % for 4 months. $ ____

10) $8,000 at 3.5% for 4 years. $ ___

Calculate each simple interest word problem.

11) A new car, valued at $18,000, depreciates at 5.5% per year. What is the value of the car one year after purchase? $_____

12) Sara puts $9,000 into an investment yielding 8% annual simple interest; she left the money in for two years. How much interest does Sara get at the end of those three years? $_____

13) A bank is offering 12.5% simple interest on a savings account. If you deposit $30,400, how much interest will you earn in four years? $_____

14) $1,200 interest is earned on a principal of $5,000 at a simple interest rate of 12% interest per year. For how many years was the principal invested? _____

15) In how many years will $1,800 yield an interest of $432 at 6% simple interest? _____

16) Jim invested $8,000 in a bond at a yearly rate of 2.5%. He earned $600 in interest. How long was the money invested? _____

Answer key Chapter 6

Proportions

1) $a = 21$

2) $a = 4$

3) $a = 1$

4) $a = 6$

5) $a = 18.75$

6) $a = 24$

7) $a = 36$

8) $a = 17.5$

9) $a = 10$

10) $a = 48$

11) $a = 3$

12) $a = 76$

13) $a = \frac{48}{11}$

14) $a = 24$

15) $a = 36$

16) $a = 120$

17) $a = 14.28$

18) $a = 1$

19) $a = 3.2$

20) $a = 21$

Reduce Ratio

1) $1 : 4$

2) $1 : 6$

3) $9 : 5$

4) $6 : 5$

5) $1 : 10$

6) $8 : 1$

7) $1 : 5$

8) $1 : 36$

9) $1 : 3$

10) $0.6 : 0.7$

11) $11 : 33$

12) $0.6 : 1$

13) $3 : 5$

14) $2 : 3$

15) $1 : 2$

16) $4 : 1$

17) $4 : 1$

18) $1 : 10$

19) $3 : 2$

20) $26 : 33$

21) $1 : 2$

22) $6 : 7$

23) $1 : 15$

24) $1 : 6$

Percent

1) 12.6

2) 4.2

3) 2.1

4) 16.8

5) 6.4

6) 4.2

7) 0.9

8) 5.52

9) 24.8

10) 2.25

11) 15.3

12) 30

13) 13.4

14) 4.8

15) 0

16) 96

17) 16.2

18) 7

19) 1.04

20) 0.32

21) 13.2

22) 13.6

23) 7.7

24) 12

25) 108

26) 0.1

27) 28.8

28) 2.6

Discount, Tax and Tip

1) $162.00	7) $322.00	13) $62.50
2) $276.00	8) $99.00	14) $84.96
3) $324.00	9) $27.00	15) $100.00
4) $13,041.00	10) $633.60	16) $86.40
5) $560.00	11) $31.05	
6) $284,200	12) $49.50	

Percent of Change

1) 100%	7) 50%	13) 20%
2) 400%	8) 50%	14) 25%
3) 600%	9) 40%	15) 50%
4) 300%	10) 75%	16) 15%
5) 200%	11) 25%	17) 20%
6) 80%	12) 60%	

Simple Interest

1) $126.00	7) $2,960.00	13) $15,200.00
2) $216.00	8) $25.00	14) 2 years
3) $900.00	9) $13.00	15) 4 years
4) $92.00	10) $1,120.00	16) 3 years
5) $17.50	11) $17,010.00	
6) $10,200.00	12) $2,160.00	

Chapter 7:
Monomials and
Polynomials

Adding and Subtracting Monomial

Simplify each expression.

1) $3x^3 + 10x^3 =$

2) $5x^2 + 2x^2 =$

3) $\frac{1}{8}x^3 + \frac{4}{8}x^3 =$

4) $3\frac{1}{4}x^4 + 5\frac{3}{4}x^4 =$

5) $11x^7 - 5x^7 =$

6) $6.9x^3 - 2.9x^3 =$

7) $-x^{10} + x^{10} =$

8) $(x^4)^5 + (x^5)^4 =$

9) $3x^{-5} + 2x^{-5} =$

10) $15p^8 - (-5p^8) =$

11) $2x^2 - 3.8x^2 =$

12) $4\frac{1}{5}x^4 + 3\frac{2}{5}x^4 =$

13) $-4\frac{1}{8}x^{13} + 6\frac{1}{8}x^{13} =$

14) $\sqrt{81}p^6 + (-5p^6) =$

15) $(-1.8p^4) + (-3.2p^4) =$

16) $-1.2x^6 + 7.4x^6 =$

17) $x^8 + \frac{2}{5}x^8 =$

18) $12x^4 - x^4 =$

19) $-3.2x^2 - 6.9x^2 =$

20) $-6x^7 - 3x^7 =$

21) $32x^5 - 20x^5 =$

22) $-3x^{13} + 7x^{13} =$

23) $5x^{-10} - 16x^{-10} =$

24) $17x^{-7} - 7x^{-7} =$

Multiplying and Dividing Monomial

Simplify.

1) $5xy^3 \times 2x^4 =$

2) $7xy \times 4x^3 =$

3) $5xy^3 \times (-6x^3y^4) =$

4) $5x^6y^{10} \times x^3y^2 =$

5) $12x^2 \times (-5x^4) =$

6) $-5x^2y^4z \times 3x^3y^3z^4 =$

7) $-7 \times (-11x^{14}y^{16}) =$

8) $3x^2y^4 \times (-12x^2y^5) =$

9) $6x^2 \times (-8x) =$

10) $-8x^3y^8 \times 3x^2y =$

11) $22x^{-4}y^6 \times (-x^{-7}y^{-3}) =$

12) $6x^{10}y^3z \times 5xy^{-3}z =$

13) $\dfrac{20x^{11}y^4}{10x^5y^2} =$

14) $(6y^5)^{-2} =$

15) $\dfrac{120x^{18}y^6}{12x^8y^3} =$

16) $\dfrac{28x^{14}}{7x^8} =$

17) $\dfrac{33x^7y^4z^3}{11x^2y^4z} =$

18) $\dfrac{15x^2+10x}{5x} =$

19) $\dfrac{200x^5y^9}{100x^5y^8} =$

20) $(12x^2)(6x^2) =$

21) $\dfrac{36x^2y^4+18xy^6}{9xy} =$

22) $\dfrac{-48x^7y^{14}}{6x^5y^{11}} =$

23) $\dfrac{15x^6y^{15}}{10x^4y^4} =$

24) $\dfrac{36x^{16}y^{12}z}{9x^4y^5} =$

Binomial Operations

Solve each operation below.

1) $3x + 8 - (6x - 3) =$

2) $(4x - 5) + (6x - 7) =$

3) $(-5x - 5) + (8x + 2) =$

4) $(4x - 1.4) + (5x - 3.6) =$

5) $\frac{1}{8}x + 4 - \left(\frac{1}{3}x - 5\right) =$

6) $5x + 3 - (7x - 2) =$

7) $14x + 5 - (26x - 1) =$

8) $(x + 8)(x + 5) =$

9) $(x - 7)(x - 3) =$

10) $(x - 4)(3x + 4) =$

11) $(x - 8)(x + 8) =$

12) $(x - 6)(9x + 5) =$

13) $(4x - 5)(4x + 5) =$

14) $(x + 9)(x - 6) =$

15) $(x - 7)(5x + 7) =$

16) $(x^2 + 6)(x^2 - 6) =$

17) $(x - 3)(x + 3) =$

18) $7x(6x - 4) =$

19) $13x(3x + 5) =$

20) $(3x + 4) + (5x - 7) + (x - 9) =$

21) $(x^2 + 2)(x^2 - 2)$

22) $(3x - 3)(5x + 4)$

23) $(x - 5)(7x + 2)$

24) $(x - 3.4)(4.1x + 3.4)$

Polynomial Operations

Simplify each expression.

1) $(4x^2 + x - 6) + (2x - 4x^2 - 8) =$

2) $(3x^2 + 2x - 3) - (4x - 3x^2 - 2) =$

3) $(12x^2 - 8x + 3) - (-4x + 6x^2 - 3) =$

4) $(6x^5 - 2x^3 - 7x) + (5x + 14x^4 - 15) + (3x^2 + x^3 + 12) =$

5) $(15x^2 - 8x + 4) + (6x^2 - 2x + 3) =$

6) $12(3x^2 - 7x - 2) =$

7) $3x^3(3x^2 - 3x + 5) =$

8) $3x^2y^2(4x^2 - 6x + 3) =$

9) $(x + 5)(x^2 - 4x + 7) =$

10) $x(3x^2 - 3x + 7) =$

11) $5(x^2 - 3x + 6) =$

12) $(x - 3)(x^2 + 7x - 2) =$

13) $(12x^3 + 5x^2 - 13) + (-5x^3 + 3x^2 + 11) =$

14) $(3x - 4)(3x^2 + 5x + 10) =$

Squaring a Binomial

Write each square as a trinomial.

1) $(a + 2b)^2 =$

2) $(x + 5)^2 =$

3) $(2a - b)^2 =$

4) $(3x + 2)^2 =$

5) $(x - 8)^2 =$

6) $(2x + \frac{1}{4})^2 =$

7) $(4x - 5y)^2 =$

8) $(x - 7)^2 =$

9) $(x + 11)^2 =$

10) $(4x - 5)^2 =$

11) $(3x + 3y)^2 =$

12) $(3x + 8)^2 =$

13) $(3x + \frac{1}{3})^2 =$

14) $(2x^2 + 2y^2)^2 =$

15) $(x - 12)^2 =$

16) $\left(x + \sqrt{3}\right)^2 =$

17) $(6x - 7)^2 =$

18) $2(x + 3)^2 =$

19) $(8x - 3y)^2 =$

20) $(x + 13)^2 =$

21) $2(x + 2)^2 =$

22) $(x^2 - 9)^2 =$

23) $(2x + 5)^2 =$

24) $(4x + 7)^2 =$

Factor polynomial

Factor each completely.

1) $x^2 + 17x + 66 =$

2) $12x^2 - 36x =$

3) $x^3 - 7x^2 - 7x + 49 =$

4) $x^2 + 12x + 32 =$

5) $x^4 - 4x^2 - 12 =$

6) $x^2 - 14x + 45 =$

7) $10 + 6x + 32 + x =$

8) $3x^2 - 18x + 16x - 4 =$

9) $24x^3y + 8x^2y - 32xy =$

10) $20x^2 - 7x - 3 =$

11) $3x^2 - 26x + 35 =$

12) $\dfrac{3x^2 - 15x + 18}{x^2 - 9x + 14} =$

13) $\dfrac{(x-3)(x-5)}{(x-3)(x-9)} =$

14) $(x - 4)4x + (x - 4)4 =$

15) $8x^2 - 20x^4 =$

16) $\dfrac{x^2 + 10x + 24}{(x+4)} =$

17) $x^2 + 2x - 63 =$

18) $4x^4 + 24x^2 - 28x^3 - 168x =$

19) $12(a - b) - 4a(a - b) =$

20) $18x^2 - 18x =$

Answer key Chapter 7

Adding and Subtracting Monomial

1) $13x^3$
2) $7x^2$
3) $\frac{5}{8}x^3$
4) $9x^4$
5) $6x^7$
6) $4x^3$
7) 0
8) $2x^{20}$

9) $5x^{-5}$
10) $20p^8$
11) $-1.8x^2$
12) $7\frac{3}{5}x^4$
13) $2x^{13}$
14) $4p^6$
15) $-5p^4$
16) $6.2x^6$

17) $\frac{7}{5}x^8$
18) $11x^4$
19) $-10.1x^2$
20) $-9x^7$
21) $12x^5$
22) $4x^{13}$
23) $-11x^{-10}$
24) $10x^{-7}$

Multiplying and Dividing Monomial

1) $10x^5y^3$
2) $28x^4y$
3) $-30x^4y^7$
4) $5x^9y^{12}$
5) $-60x^6$
6) $-15x^5y^7z^5$
7) $77x^{14}y^{16}$
8) $-36x^4y^9$
9) $-48x^3$

10) $-24x^5y^9$
11) $-22x^{-11}y^3$
12) $30x^{11}z^2$
13) $2x^6y^2$
14) $\frac{1}{36}y^{-10}$
15) $10x^{10}y^3$
16) $4x^6$
17) $3x^5z^4$

18) $3x + 2$
19) $2y$
20) $72x^4$
21) $4xy^3 + 2y^5$
22) $-8x^2y^3$
23) $\frac{3}{2}x^2y^{11}$
24) $4x^{12}y^7z$

Binomial Operations

1) $-3x + 11$
2) $10x - 12$
3) $3x - 3$
4) $9x - 5$
5) $-\frac{5}{24}x + 9$
6) $-2x + 5$
7) $-12x + 6$
8) $x^2 + 13x + 40$

9) $x^2 - 10x + 21$
10) $3x^2 - 8x - 16$
11) $x^2 - 64$
12) $9x^2 - 49x - 30$
13) $16x^2 - 25$
14) $x^2 + 3x - 54$
15) $5x^2 - 28x - 49$
16) $x^4 - 36$

17) $x^2 - 9$
18) $42x^2 - 28x$
19) $39x^2 + 65x$
20) $9x - 12$
21) $x^4 - 4$
22) $15x^2 - 3x - 12$
23) $7x^2 - 33x - 10$
24) $4.1x^2 - 10.54x - 11.56$

Polynomial Operations

1) $3x - 14$

2) $6x^2 - 2x - 1$

3) $6x^2 - 4x + 6$

4) $6x^5 + 14x^4 - x^3 + 3x^2 - 3$

5) $21x^2 - 10x + 7$

6) $36x^2 - 84x - 24$

7) $9x^5 - 9x^4 + 15x^3$

8) $12x^4y^2 - 18x^3y^2 + 9x^2y^2$

9) $x^3 + x^2 - 13x + 35$

10) $3x^3 - 3x^2 + 7x$

11) $5x^2 - 15x + 30$

12) $x^3 + 4x^2 - 23x + 6$

13) $7x^3 + 8x^2 - 2$

14) $9x^3 + 3x^2 + 10x - 40$

Squaring a Binomial

1) $a^2 + 4b^2 + 4ab$

2) $x^2 + 10x + 25$

3) $4a^2 + b^2 - 4ab$

4) $9x^2 + 12x + 4$

5) $x^2 - 16x + 64$

6) $x^2 + x + \frac{1}{16}$

7) $16x^2 - 40xy + 25y^2$

8) $x^2 - 14x + 49$

9) $x^2 + 22x + 121$

10) $16x^2 - 40x + 25$

11) $9x^2 + 18xy + 9y^2$

12) $9x^2 + 48x + 64$

13) $9x^2 + 2x + \frac{1}{9}$

14) $4x^4 + 4y^4 + 8x^2y^2$

15) $x^2 - 24x + 144$

16) $x^2 + 2\sqrt{3}x + 3$

17) $36x^2 - 84x + 49$

18) $2x^2 + 12x + 18$

19) $64x^2 - 48xy + 9y^2$

20) $x^2 + 36x + 169$

21) $2x^2 + 8x + 8$

22) $x^4 - 18x^2 + 81$

23) $4x^2 + 20x + 25$

24) $16x^2 + 56x + 49$

Factor polynomial

1) $(x + 6)(x + 11)$

2) $12x(x - 3)$

3) $(x^2 - 7)(x - 7)$

4) $(x + 8)(x + 4)$

5) $(x^2 - 6)(x^2 + 2)$

6) $(x - 9)(x - 5)$

7) $7(x + 6)$

8) $3x(x - 2) + 4(x - 1)$

9) $8xy(3x^2 + x - 4)$

10) $(4x + 1)(5x - 3)$

11) $(3x - 5)(x - 7)$

12) $\frac{3(x-3)}{x-7}$

13) $\frac{x-5}{x-9}$

14) $(x - 4)(4x + 4)$

15) $-2x^2(-4 + 10x^2)$

16) $x + 6$

17) $(x + 9)(x - 7)$

18) $4x(x^2 + 6)(x - 7)$

19) $(a - b)(12 - 4a)$

20) $18x(x - 1)$

Chapter 8:
Functions

Relation and Functions

Determine whether each relation is a function. Then state the domain and range of each relation.

1)

Function:

....

Domain:

....

Range:

....

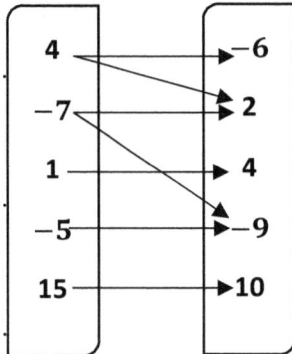

2)

Function:

....

Domain:

....

Range:

....

x	y
4	5
2	3
−6	−8
6	−8
−11	2

3)

Function:

....

Domain:

....

Range:

....

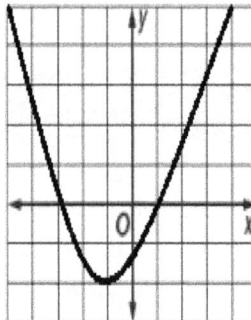

4) $\{(2, -2), (7, -6), (9, 9), (8, 1), (7, 4)\}$

Function:

....

Domain:

....

Range:

....

5)

Function:

....

Domain:

....

Range:

....

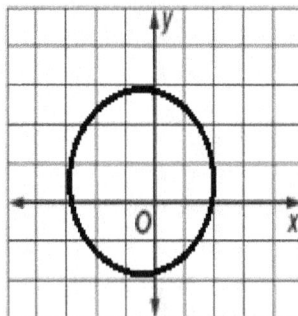

6)

Function:

....

Domain:

....

Range:

....

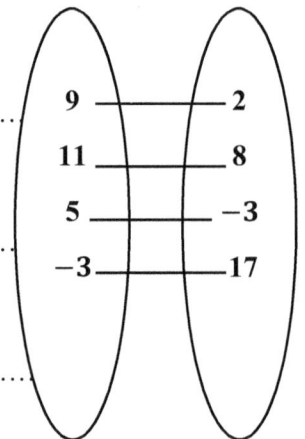

Slope form

Write the slope-intercept form of the equation of each line.

1) $4x + 5y = 15$

2) $4x + 12y = 3$

3) $7x + y = -9$

4) $-7x + 11y = 5$

5) $5x - 4y = 7$

6) $-21x + 3y = 6$

7) $2x + y = 0$

8) $5x - 7y = -9$

9) $-13.5x + 27y = 54$

10) $-3x + \frac{2}{3}y = 18$

11) $10x + y = -120$

12) $3x = -36y - 27$

13) $1.5x = 3y + 3$

14) $5x = -\frac{5}{4}y + 25$

Slope and Y-Intercept

Find the slope and y-intercept of each equation.

1) $y = \frac{1}{5}x + 4$

2) $y = 7x + 8$

3) $x - 3y = 9$

4) $y = 5x + 21$

5) $y = 9$

6) $y = -2x + 3$

7) $x = -15$

8) $y = 7x$

9) $y - 3 = 4(x + 1)$

10) $x = -\frac{5}{8}y - \frac{1}{3}$

Slope and One Point

Find a Point-Slope equation for a line containing the given point and having the given slope.

1) $m = -2, (1, -1)$

2) $m = 3, (1, 2)$

3) $m = -2, (-1, -5)$

4) $m = 1, (3, 2)$

5) $m = 5, (2, 4)$

6) $m = \frac{3}{2}, (4, 5)$

7) $m = 0, (-4, -5)$

8) $m = 2, (1, -3)$

9) $m = 1, (0, 3)$

10) $m = \frac{3}{4}, (-2, -5)$

11) $m = -3, (1, -1)$

12) $m = -2, (2, -1)$

13) $m = 5, (1, 0)$

14) $m =$ undefined, $(8, -8)$

15) $m = -\frac{1}{8}, (8, 4)$

16) $m = \frac{1}{4}, (3, 2)$

17) $m = -8, (2, 4)$

18) $m = 6, (-2, -4)$

19) $m = \frac{1}{3}, (3, 1)$

20) $m = \frac{-4}{9}, (0, -3)$

21) $m = \frac{1}{4}, (4, 4)$

22) $m = -5, (0, -1)$

23) $m = 0, (0.9, -3)$

24) $m = -\frac{5}{7}, (7, -1)$

25) $m = 0, (-4, 8)$

26) $m =$ Undefined, $(-10, -2)$

Slope of Two Points

Write the slope-intercept form of the equation of the line through the given points.

1) $(1, 0), (-1, 5)$

2) $(-1, 3), (5, 6)$

3) $(-5, 1), (-1, 5)$

4) $(2, -3), (-9, 8)$

5) $(5, 0), (3, 1)$

6) $(9, -1), (-1, 9)$

7) $(-5, 3), (-6, 1)$

8) $(-7, -2), (1, 0)$

9) $(-5, -5), (3, 3)$

10) $(-1, 9), (-1, -5)$

11) $(-2, 7), (1, 7)$

12) $(1, -5), (4, -4)$

13) $(6, -9), (-3, 0)$

14) $(1, -4), (7, 4)$

15) $(-9, 5), (-3, -1)$

16) $(9, 5), (5, 1)$

17) $(10, -7), (2, -6)$

18) $(-5, -9), (-7, 2)$

19) $(7, 4), (3, 1)$

20) $(-1, -1), (9, 2)$

21) $(-8, 8), (8, 2)$

22) $(9, 2), (5, 11)$

23) $(8, 2), (9, 3)$

24) $(-2, -5), (-5, -8)$

Equation of Parallel and Perpendicular lines

Write the slope-intercept form of the equation of the line described.

1) Through: $(-5, 2)$, parallel to $y = 2x + 5$

2) Through: $(-4, 1)$, parallel to $y = -3x$

3) Through: $(-10, -2)$, perpendecular to $y = \frac{1}{2}x + 8$

4) Through: $(6, -2)$, parallel to $y = -5x + 13$

5) Through: $(-7, 4)$, parallel to $y = \frac{3}{7}x - 6$

6) Through: $(2, 0)$, perpendecular to $y = -\frac{1}{5}x + 8$

7) Through: $(4, -7)$, perpendecular to $y = -6x - 10$

8) Through: $(-5, 1)$, perpendecular to $y = -\frac{1}{8}x + 3$

9) Through: $(-1, -2)$, parallel to $2y + 4x = 9$

10) Through: $(1, 10)$, parallel to $y = \frac{1}{10}x - 5$

11) Through: $(5, -5)$, parallel to $y = 9$

12) Through: $(7, 2)$, perpendecular to $y = \frac{5}{2}x + 3$

13) Through: $(0, -4)$, perpendecular to $3y - x = 11$

14) Through: $(3, 5)$, parallel to $3y + x = 5\frac{3}{4}$

15) Through: $(1, 1)$, perpendecular to $y = 5x + 12$

16) Through: $(-3, -5)$, parallel to $8y - x = 10$

17) Through: $(-2, -2)$, perpendecular to $y = 4x + \frac{1}{7}$

18) Through: $(-8, 0)$, perpendecular to $5y - 4x - 9 = 0$

Quadratic Equations - Square Roots Law

Solve each equation by taking square roots.

1) $x^3 - 4 = 4$

2) $x^2 - 6 = 19$

3) $12x^2 - 2 = 100$

4) $-x^2 - 2 = -66$

5) $6x^2 + 3 = 489$

6) $5x^2 + 9 = 14$

7) $8x^2 - 17 = 2,791$

8) $7x^2 + 16 = 2,151$

9) $100x^2 = 4$

10) $4x^2 - 8 = 68$

11) $9x^2 - 5 = 607$

12) $20x^2 - 20 = 60$

13) $13x^2 - 3 = 4,209$

14) $13x^2 - 8 = -1,139$

15) $16x^2 - 16 = 48$

16) $54x^2 = 6$

17) $-8x^2 - 8 = -31$

18) $15x^2 - 30 = 255$

19) $42x^2 = -126$

20) $16x^2 - 10 = 39$

21) $2x^2 + 16 = 160$

22) $13x^2 = 117$

23) $-16x^2 + 12 = 340$

24) $6x^2 - 30 = -24$

25) $12x^2 - 60 = -48$

26) $3x^2 + 15 = 315$

27) $27x^2 + 3 = 975$

28) $21x^2 + 3 = 87$

Quadratic Equations - Factoring

Solve each equation by factoring.

1) $(10n - 5)(9n + 3) = 0$

2) $(20n - 4)(4n + 4) = 0$

3) $x^2 + 16 = 10x$

4) $7x^2 - 42 = -35x$

5) $2x^2 - 7 = 13x$

6) $7x^2 + 32 = 7 - 40x$

7) $-6x^2 + 2x + 144 = 6x^2 + 14x$

8) $8x^2 + 3x + 2 = 7x^2$

9) $8x^2 - 57x = -7$

10) $21x^2 + 75 = -120x$

11) $3n(2n - 10) = 0$

12) $(n + 4)(5n - 6) = 0$

13) $x^2 + 9 = -9 - 9x$

14) $5x^2 + 12 = -20x - 8$

15) $3x^2 + 20 = -18x - 7$

16) $5x^2 + 2x = 10 - 3x$

17) $3x^2 - 8x + 4 = 20$

18) $2x^2 + 40 = -24x - 32$

19) $3x^2 - 8x = 16$

20) $3x^2 + 50 = -30x - 25$

21) $6x^2 - 16x - 32 = 0$

22) $5x^2 + 5x - 50 = 10$

23) $2x^2 + 6x = -10x - 32$

24) $2n(n + 2) = 0$

Quadratic Equations - Completing the Square

Solve each equation by completing the square.

1) $3x^2 - 6x - 9 = 0$

2) $5x^2 - 10x - 15 = 0$

3) $2x^2 + \frac{5x}{2} - 3 = 0$

4) $-3x^2 + 2x + 8 = 0$

5) $3x^2 + 42x - 153 = 0$

6) $3x^2 + 18x + 4 = -20$

7) $x^2 + 14x - 2 = 13$

8) $x^2 - 4x - 91 = 7$

9) $2x^2 - 18x = -36$

10) $5x^2 = -20x + 60$

11) $12x^2 = 48x - 36$

12) $\frac{1}{4}x^2 - \frac{2}{4}x - \frac{3}{4} = 0$

13) $3x^2 - 36x + 25 = -8$

14) $3x^2 - 30x + 54 = 0$

15) $x^2 + 6x = 59$

16) $3x^2 = 54x + 120$

17) $\frac{1}{5}x^2 + \frac{2}{5}x = -4$

18) $2x^2 - 24x = 22$

19) $6x^2 - 12x - 18 = 0$

20) $3x^2 + 42x - 45 = 0$

21) $\frac{1}{2}x^2 - x - \frac{3}{2} = 0$

22) $7x^2 - 28x = -21$

23) $9x^2 + 54x + 72 = 0$

24) $-7x^2 - 42x = 56$

Quadratic Equations - Quadratic Formula

Solve each equation with the quadratic formula.

1) $3x^2 + 6x - 24 = 0$

2) $4x^2 - 8x = -4$

3) $\frac{1}{4}x^2 = \frac{9}{4}x - 5$

4) $\frac{2}{3}x^2 + \frac{10}{3}x - 4 = 0$

5) $3x^2 = 27x - 60$

6) $3x^2 - 12x - 26 = 10$

7) $7x^2 = -21x + 280$

8) $3x^2 + 15x - 18 = 0$

9) $x^2 + x - 2 = \frac{1}{4}$

10) $\frac{4}{3}x^2 - \frac{2}{3}x = 3$

11) $x^2 = -3x + 40$

12) $9x^2 - 25 = 8x$

13) $\frac{8}{5}x^2 - \frac{8}{5}x = 3$

14) $6x^2 - 3x - 10 = 8$

15) $\frac{1}{3}x^2 = 3x - \frac{100}{15}$

16) $10x^2 + 30x = 400$

17) $x^2 - 2 = \frac{1}{8}x$

18) $24x^2 + 18x + 15 = 0$

19) $x^2 - \frac{1}{2}x - \frac{13}{2} = 1$

20) $11x^2 + 1 = 5x^2 + 7x$

21) $17x^2 + 14 = x$

22) $10x^2 - 5x - 20 = 10$

Arithmetic Sequences

Find the three terms in the sequence after the last one given.

1) $2, 6, 10, 14, ..., ..., ...,$

2) $-10, -6, -2, 2, ..., ..., ...,$

3) $a_1 = -12, d = 12$

4) $a_1 = -10.6, d = 1.4$

5) $a_{10} = 36, d = 6$

6) $a_n = (3n)^2$

7) $24, 16, 8, 0, ..., ..., ...,$

8) $-15, -30, -45, -60, ..., ..., ...,$

9) $a_n = -18 + 3n$, find a_{26}

10) $a_n = -8.4 - 5.8n$, find a_{32}

11) $a_5 = \frac{3}{8}, d = -\frac{1}{4}$

12) $a_n = \frac{2n^2}{3n+2}$

13) $-8, -15, -22, ..., ..., ...,$

14) $a_n = 4n + 3$

15) $-8, -3, 2, 7, ..., ..., ...,$

16) $-12, -7, -2, 3, ..., ..., ...,$

17) $\frac{6}{7}, \frac{5}{14}, -\frac{1}{7}, -\frac{9}{14}, ..., ..., ...,$

18) $5, -5, -15, ..., ..., ...,$

19) $9, 13, 17, ..., ..., ...,$

20) $11, 22, 33, ..., ..., ...,$

21) $-3.5, -4.7, -5.9, ..., ..., ...,$

22) $115, 140, 165, 190, ..., ..., ...,$

23) $-\frac{8}{9}, -\frac{5}{9}, -\frac{2}{9}, ..., ..., ...,$

24) $-2.25, -1.50, -0.75, ..., ..., ...,$

Geometric Sequences

Find the three terms in the sequence after the last one given.

1) $3, 9, 27, 81, \ldots, \ldots, \ldots$

2) $320, 80, 20, \ldots, \ldots, \ldots$

3) $a_n = a_{n-1} \times (-2), a_1 = 1$

4) $a_n = a_{n-1} \times 4, a_1 = 3$

5) $0.2, 0.6, 1.8, 5.4 \ldots, \ldots, \ldots$

6) $a_n = 2^{n-1}, a_1 = 1$

7) $a_n = 8(\frac{1}{4})^{n-1}, a_1 = 8$

8) $1, \frac{3}{2}, \frac{9}{4}, \frac{27}{8}, \ldots \ldots, \ldots \ldots, \ldots$

9) $a_n = 2a_{n-1}, a_1 = 5$

10) $1, 2, 4, \ldots, \ldots, \ldots$

11) $-4, -12, -36, \ldots, \ldots, \ldots$

12) $-0.2, 1, -5, \ldots, \ldots, \ldots$

13) $a_n = (-2)^{2n+1}$

14) $a_n = 0.25 \times (4)^{n-1} \ find \ a_5$

15) $a_n = 3 \times (2)^{n-1} ; a_6 = ?$

16) $3, 1, \frac{1}{3}, \ldots, \ldots, \ldots, \ldots$

17) $2, 6, 18, \ldots, \ldots, \ldots$

18) $5, -5, 5, \ldots, \ldots, \ldots$

19) $243, 162, 108, \ldots, \ldots, \ldots$

20) $-1.5, 3, -6, \ldots, \ldots, \ldots$

21) $-0.25, -1, -4, \ldots, \ldots, \ldots$

22) $1, -6, 36, \ldots, \ldots, \ldots$

23) $a_n = -0.1(-2)^{n-1}$

24) $a_n = -0.5 \times 4^{n-1}$

Answer key Chapter 8

Relation and Functions

1) No, $D_f = \{4, -7, 1, -5, 15\}$, $R_f = \{-6, 2, 4, -9, 10\}$

2) Yes, $D_f = \{4, 2, -6, 6, -11\}$, $R_f = \{5, 3, -8, 2\}$

3) Yes, $D_f = (-\infty, \infty)$, $R_f = \{-2, \infty\}$

4) No, $D_f = \{2, 7, 9, 8, 7\}$, $R_f = \{-2, -6, 9, 1, 4\}$

5) No, $D_f = [-3, 2]$, $R_f = [-2, 3]$

6) Yes, $D_f = \{9, 11, 5, -3\}$, $R_f = \{2, 8, -3, 17\}$

Slope form

1) $y = -\frac{4}{5}x + 3$

2) $y = -\frac{1}{3}x + \frac{1}{4}$

3) $y = -7x - 9$

4) $y = \frac{7}{11}x + \frac{5}{11}$

5) $y = \frac{5}{4}x - \frac{7}{4}$

6) $y = 7x + 2$

7) $y = -2x$

8) $y = \frac{5}{7}x + \frac{9}{7}$

9) $y = 0.5x + 2$

10) $y = 4.5x + 27$

11) $y = -10x - 120$

12) $y = -\frac{1}{12}x - \frac{3}{4}$

13) $y = 0.5x - 1$

14)

Slope and Y-Intercept

1) $m = \frac{1}{5}, b = 4$

2) $m = 7, b = 8$

3) $m = \frac{1}{3}, b = -3$

4) $m = 5, b = 21$

5) $m = 0, b = 9$

6) $m = -2, b = 3$

7) $m = undefind,$
$b: no\ intercept$

8) $m = 7, b = 0$

9) $m = 4, b = 7$

10) $m = -\frac{8}{5}, b = -\frac{1}{3}$

Slope and One Point

1) $y = -2x + 1$

2) $y = 3x - 1$

3) $y = -2x - 7$

4) $y = x - 1$

5) $y = 5x - 6$

6) $y = \frac{3}{2}x - 1$

7) $y = -5$

8) $y = 2x - 5$

9) $y = x + 3$

10) $y = \frac{3}{4}x - \frac{7}{2}$

11) $y = -3x + 2$

12) $y = -2x + 3$

13) $y = 5x$

14) $x = 8$

15) $y = -\frac{1}{8}x + 5$

16) $y = \frac{1}{4}x + \frac{5}{4}$

17) $y = -8x + 20$

18) $y = 6x + 8$

19) $y = \frac{1}{3}x$

20) $y = -\frac{4}{9}x - 3$

21) $y = \frac{1}{4}x + 3$

22) $y = -5x - 1$

23) $y = -3$

24) $y = -\frac{5}{7}x + 4$

25) $y = 8$

26) $x = -10$

Slope of Two Points

1) $y = -\frac{5}{2}x + \frac{5}{2}$

2) $y = \frac{1}{2}x + \frac{7}{2}$

3) $y = x + 6$

4) $y = -x - 1$

5) $y = -\frac{1}{2}x + \frac{5}{2}$

6) $y = -x + 8$

7) $y = 2x + 13$

8) $y = \frac{1}{4}x - \frac{1}{4}$

9) $y = x$

10) $x = -1$

11) $y = 7$

12) $y = \frac{1}{3}x - 5\frac{1}{3}$

13) $y = -x - 3$

14) $y = \frac{4}{3}x - 5\frac{1}{3}$

15) $y = -x - 4$

16) $y = x - 4$

17) $y = -\frac{1}{8}x - 5\frac{3}{4}$

18) $y = -5\frac{1}{2}x - 36\frac{1}{2}$

19) $y = \frac{3}{4}x - 1\frac{1}{4}$

20) $y = \frac{3}{10}x - \frac{7}{10}$

21) $y = -\frac{3}{8}x + 5$

22) $y = -\frac{9}{4}x + 22\frac{1}{4}$

23) $y = x - 6$

24) $y = x - 3$

Equation of Parallel and Perpendicular lines

1) $y = 2x + 12$

2) $y = -3x - 11$

3) $y = -2x - 22$

4) $y = -5x + 28$

5) $y = \frac{3}{7}x + 7$

6) $y = 5x - 10$

7) $y = \frac{1}{6}x - 7\frac{2}{3}$

8) $y = 8x + 41$

9) $y = -2x - 4$

10) $y = \frac{1}{10}x + 9\frac{9}{10}$

11) $y = -5$

12) $y = -\frac{2}{5}x + 4\frac{4}{5}$

13) $y = -3x - 4$

14) $y = -\frac{1}{3}x + 6$

15) $y = -\frac{1}{5}x + 1\frac{1}{5}$

16) $y = \frac{1}{8}x - 4\frac{5}{8}$

17) $y = -\frac{1}{4}x - 2\frac{1}{2}$

18) $y = -\frac{5}{4}x - 10$

Quadratic Equations - Square Roots Law

1) 2

2) $\{5, -5\}$

3) $\{\frac{\sqrt{17}}{2}, -\frac{\sqrt{17}}{2}\}$

4) $\{8, -8\}$

5) $\{9, -9\}$

6) $\{1, -1\}$

7) $\{3\sqrt{39}, -3\sqrt{39}\}$

8) $\{\sqrt{305}, -\sqrt{305}\}$

9) $\{\frac{1}{5}, -\frac{1}{5}\}$

10) $\{\sqrt{19}, -\sqrt{19}\}$

11) $\{2\sqrt{17}, -2\sqrt{17}\}$

12) $\{2, -2\}$

13) $\{18, -18\}$

14) $\{i\sqrt{87}, -i\sqrt{87}\}$

15) $\{2, -2\}$

16) $\{\frac{1}{3}, -\frac{1}{3}\}$

17) $\{\frac{\sqrt{46}}{4}, -\frac{\sqrt{46}}{4}\}$

18) $\{\sqrt{19}, -\sqrt{19}\}$

19) $\{i\sqrt{3}, -i\sqrt{3}\}$

20) $\{\frac{7}{4}, -\frac{7}{4}\}$

21) $\{6\sqrt{2}, -6\sqrt{2}\}$

22) $\{3, -3\}$

23) $\{i\sqrt{\frac{41}{2}}, -i\sqrt{\frac{41}{2}}\}$

24) $\{1, -1\}$

25) $\{1, -1\}$

26) $\{10, -10\}$

27) $\{6, -6\}$

28) $\{2, -2\}$

Quadratic Equations - Factoring

1) $\{\frac{1}{2}, -\frac{1}{3}\}$

2) $\{\frac{1}{5}, -1\}$

3) $\{2, 8\}$

4) $\{-6, 1\}$

5) $\{-\frac{1}{2}, 7\}$

6) $\{-\frac{5}{7}, -5\}$

7) $\{3, -4\}$

8) $\{-2, -1\}$

9) $\{\frac{1}{8}, 7\}$

10) $\{-\frac{5}{7}, -5\}$

11) $\{5, 0\}$

12) $\{-4, \frac{6}{5}\}$

13) $\{-6, -3\}$

14) $\{-2\}$

15) $\{-3\}$

16) $\{-2, 1\}$

17) $\{-\frac{4}{3}, 4\}$

18) $\{-6\}$

19) $\{-\frac{4}{3}, 4\}$

20) $\{-5\}$

21) $\{-\frac{4}{3}, 4\}$

22) $\{-4, 3\}$

23) $\{-4\}$

24) $\{-2, 0\}$

Quadratic Equations - Completing the Square

25) $\{-1, 3\}$

26) $\{3, -1\}$

27) $\{-2, \frac{3}{4}\}$

28) $\{2, -\frac{4}{3}\}$

29) $\{-17, 3\}$

30) $\{-2, -4\}$

31) $\{-15, 1\}$

32) $\{2+\sqrt{102}, 2-\sqrt{102}\}$

33) $\{3, 6\}$

34) $\{-6, 2\}$

35) $\{3, 1\}$

36) $\{-1, 3\}$

37) $\{11, 1\}$

38) $\{5+\sqrt{7}, 5-\sqrt{7}\}$

39) $\{-3+2\sqrt{17}, -3-2\sqrt{17}\}$

40) $\{-2, 20\}$

41) $\{-1+i\sqrt{19}, -1-i\sqrt{19}\}$

42) $\{6+\sqrt{47}, 6-\sqrt{47}\}$

43) $\{3, -1\}$

44) $\{-15, 1\}$

45) $\{-1, 3\}$

46) $\{3, 1\}$

47) $\{-2, -4\}$

48) $\{-2, -4\}$

Quadratic Equations - Quadratic Formula

1) $\{2, -4\}$

2) $\{1\}$

3) $\{5, 4\}$

4) $\{1, -6\}$

5) $\{5, 4\}$

6) $\{6, -2\}$

7) $\{5, -8\}$

8) $\{1, -6\}$

9) $\{\frac{-1+\sqrt{10}}{2}, \frac{-1-\sqrt{10}}{2}\}$

10) $\{\frac{1+\sqrt{37}}{4}, \frac{1-\sqrt{37}}{4}\}$

11) $\{5, -8\}$

12) $\{\frac{4+\sqrt{241}}{9}, \frac{4-\sqrt{241}}{9}\}$

13) $\{\frac{2+\sqrt{34}}{4}, \frac{2-\sqrt{34}}{4}\}$

14) $\{2, -\frac{3}{2}\}$

15) $\{5, 4\}$

16) $\{5, -8\}$

17) $\{ \frac{1+3\sqrt{57}}{16}, \frac{1-3\sqrt{57}}{16} \}$ 19) $\{3, -\frac{5}{2}\}$ 21) $\{ \frac{1+i\sqrt{951}}{34}, \frac{1-i\sqrt{951}}{34} \}$

18) $\{ \frac{-3+i\sqrt{31}}{8}, \frac{-3-i\sqrt{31}}{8}\}$ 20) $\{ 1, \frac{1}{6} \}$ 22) $\{2, -\frac{3}{2}\}$

Arithmetic sequences

1) $2, 6, 10, 14, 18, 22, 26$

2) $-10, -6, -2, 2, 6, 10, 14$

3) $-12, 0, 12, 24$

4) $-10.6, -9.2, -7.8, -6.4$

5) $-18, -12, -6, 0$

6) $9, 36, 81, 144$

7) $24, 16, 8, 0, -8, -16, -24$

8) $-15, -30, -45, -60, -75, -90, -105$

9) 60

10) -194

11) $\frac{11}{8}, \frac{9}{8}, \frac{7}{8}, \frac{5}{8}$

12) $\frac{2}{5}, 1, \frac{18}{11}, \frac{32}{14}$

13) $-8, -15, -22, -29, -36, -43$

14) $7, 11, 15, 19$

15) $-8, -3, 2, 7, 12, 17, 22$

16) $-12, -7, -2, 3, 8, 13, 18$

17) $\frac{6}{7}, \frac{5}{14}, -\frac{1}{7}, -\frac{9}{14}, -\frac{8}{7}, -\frac{23}{14}, -\frac{15}{7}$

18) $5, -5, -15, -25, -35, -45$

19) $9, 13, 17, 21, 25, 29$

20) $11, 22, 33, 44, 55, 66$

21) $-3.5, -4.7, -5.9 - 7.1, -8.3, -9.5$

22) $115, 140, 165, 190, 215, 240, 265$

23) $-\frac{8}{9}, -\frac{5}{9}, -\frac{2}{9}, \frac{1}{9}, \frac{4}{9}, \frac{7}{9}$

24) $-2.25, -1.50, -0.75, 0, 0.75, 1.5$

Geometric sequences

1) $3, 9, 27, 81, 243, 729, 2187$

2) $320, 80, 20, 5, 1.25, 0.3125$

3) $1, -2, 4, -8$

4) $3, 12, 48, 192$

5) $0.2, 0.6, 1.8, 5.4, 16.2, 48.6, 145.8$

6) $1, 2, 4, 8$

7) $8, 2, \frac{1}{2}, \frac{1}{8}$

8) $\frac{81}{16}, \frac{243}{32}, \frac{729}{64}$

9) $5, 10, 20, 40$

10) $1, 2, 4, 8, 16, 32$

11) $-4, -12, -36, -108, -324, -972$

12) $-0.2, 1, -5, 25, -125, 625$

13) $-8, -32, -128$

14) 64

15) 96

16) $3, 1, \frac{1}{3}, \frac{1}{9}, \frac{1}{27}, \frac{1}{81}$

17) $2, 6, 18, 54, 162, 486$

18) $5, -5, 5, -5, 5, -5$

19) $243, 162, 108, 72, 48, 32$

20) $-1.5, 3, -6, 12, -24, 48$

21) $-0.25, -1, -4, -16, -64, 256$

22) $1, -6, 36, -216, 1296, -7776$

23) $-0.1, \ 0.2, -0.4, 0.8$

24) $-0.5, -2, -8, -32$

Chapter 9:

Geometry

Area and Perimeter of Square

Find the perimeter and area of each squares.

1)

Perimeter:⋯⋯⋯⋯⋯

Area:⋯⋯⋯⋯⋯

2)

Perimeter:⋯⋯⋯⋯⋯

Area:⋯⋯⋯⋯⋯

3)

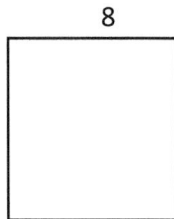

Perimeter:⋯⋯⋯⋯⋯

Area:⋯⋯⋯⋯⋯

4)

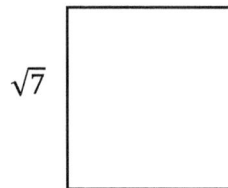

Perimeter:⋯⋯⋯⋯⋯

Area:⋯⋯⋯⋯⋯

5)

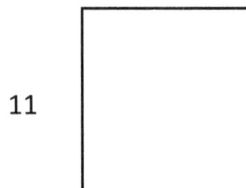

Perimeter:⋯⋯⋯⋯⋯

Area:⋯⋯⋯⋯⋯

6)

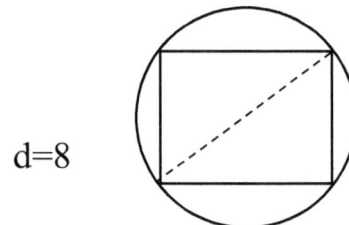

Perimeter of Square:⋯⋯⋯⋯⋯

Area of Square:⋯⋯⋯⋯⋯

Area and Perimeter of Rectangle

Find the perimeter and area of each rectangle.

1)

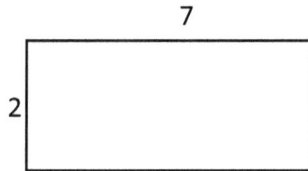

Perimeter:................:

Area:................:

2)

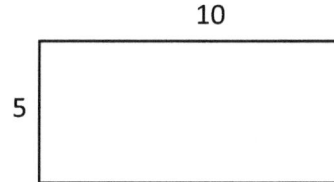

Perimeter:................:

Area:................:

3)

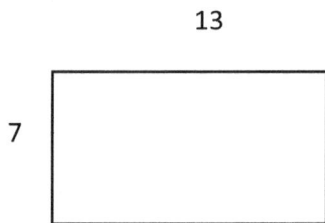

Perimeter:................:

Area:................:

4)

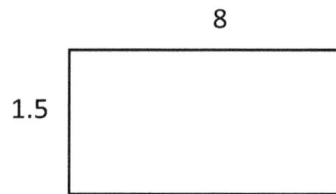

Perimeter:................:

Area:................:

5)

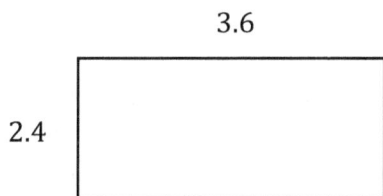

Perimeter:................:

Area:................:

6)

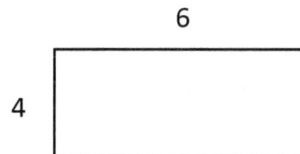

Perimeter:................:

Area:................:

Area and Perimeter of Triangle

Find the perimeter and area of each triangle.

1)

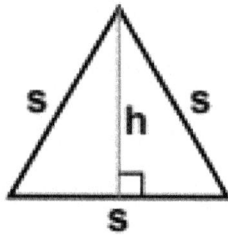

Perimeter:............:

Area:..............:

2)

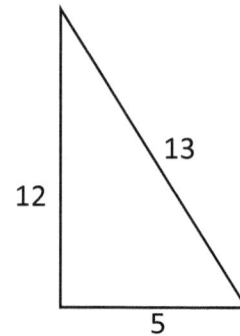

Perimeter:................:

Area:................:

3)

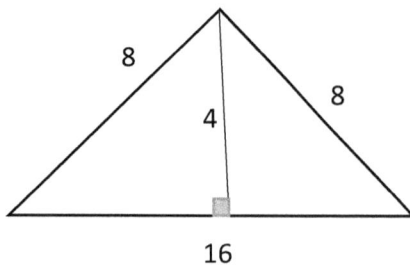

Perimeter:................:

Area   ....:

4)

s=12

h=8

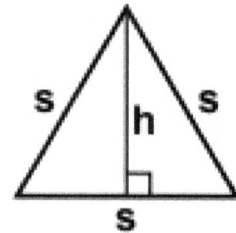

Perimeter:................:

Area:..............:

5)

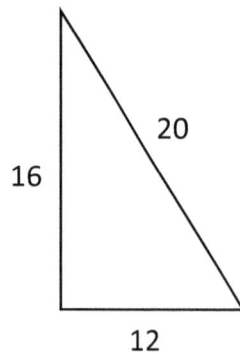

Perimeter:................:

Area:................:

6)

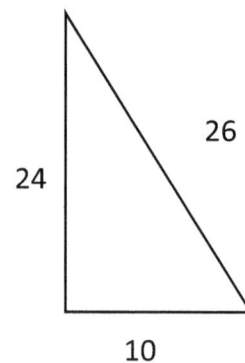

Perimeter:................:

Area:................:

Area and Perimeter of Trapezoid

Find the perimeter and area of each trapezoid.

1)

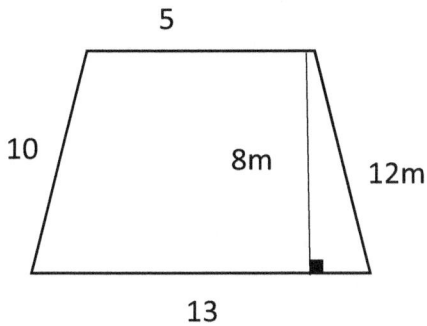

Perimeter:............:

Area:...............:

2)

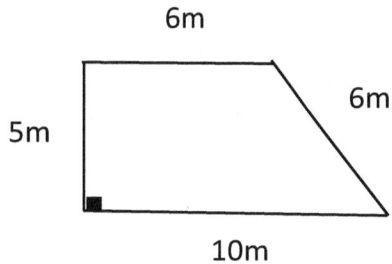

Perimeter:................:

Area:................:

3)

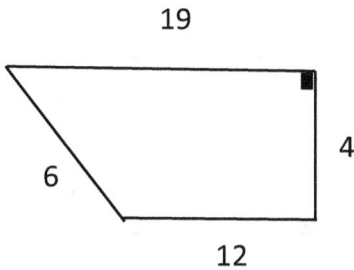

Perimeter:................:

Area ....:

4)

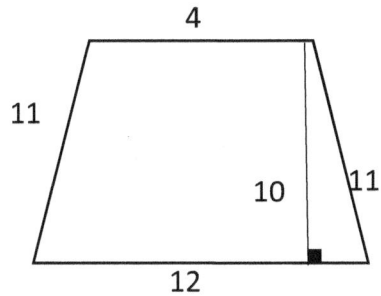

Perimeter:................:

Area:..............:

5)

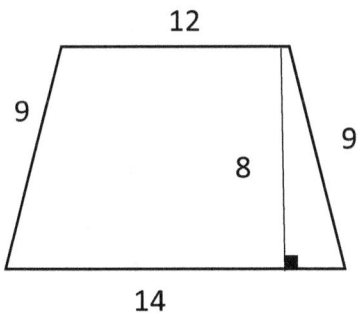

Perimeter:................:

Area:............:

6)

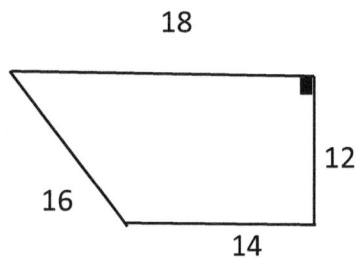

Perimeter:................:

Area:.............:

Area and Perimeter of Parallelogram

Find the perimeter and area of each parallelogram.

1)

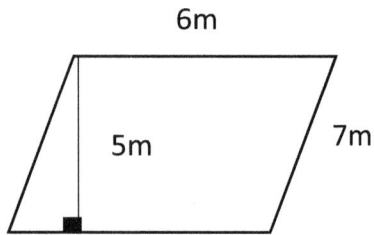

Perimeter:_____.

Area:_____.

2)

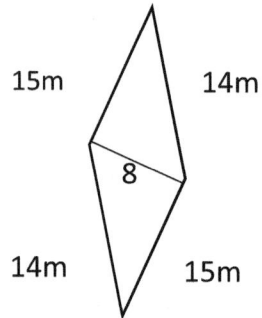

Perimeter:_____.

Area:_____.

3)

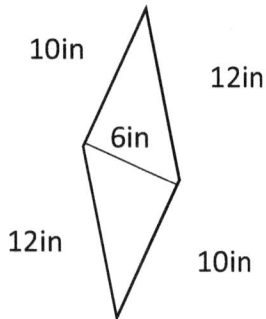

Perimeter:_____.

Area _____.

4)

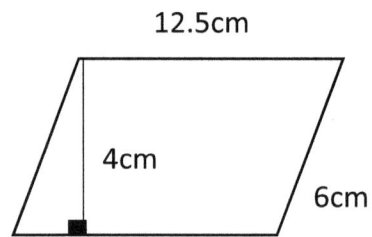

Perimeter:_____.

Area:_____.

5)

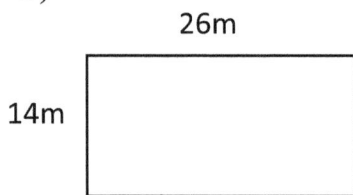

Perimeter:_____.

Area:_____.

6)

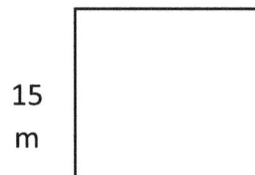

Perimeter:_____.

Area:_____.

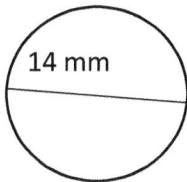

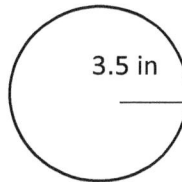

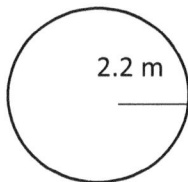

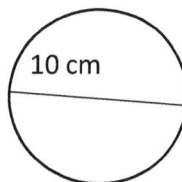

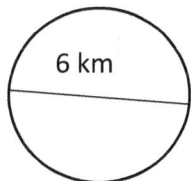

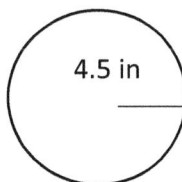

Circumference and Area of Circle

Find the circumference and area of each ($\pi = 3.14$).

1)

14 mm

Circumference:

Area:

2)

3.5 in

Circumference:

Area:

3)

2.2 m

Circumference:

Area

4)

10 cm

Circumference:

Area:

5)

6 km

Circumference:

Area:

6)

4.5 in

Circumference:

Area:

Perimeter of Polygon

Find the perimeter of each polygon.

1)

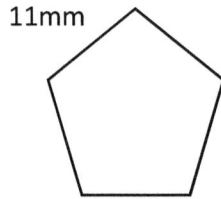

11mm

Perimeter:_____:

2)

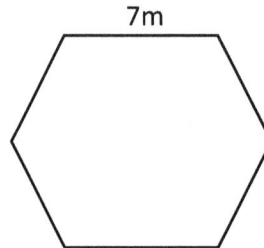

7m

Perimeter:_____:

3)

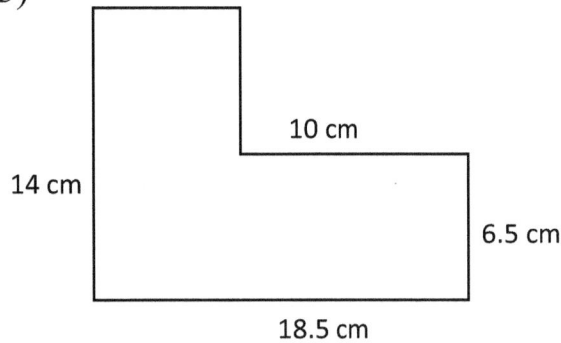

10 cm

14 cm

6.5 cm

18.5 cm

Perimeter:_____.

4)

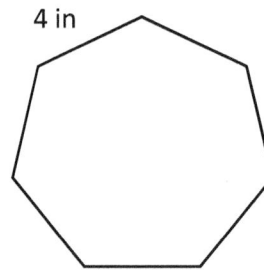

4 in

Perimeter:_____:

5)

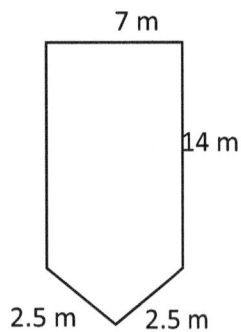

7 m

14 m

2.5 m   2.5 m

Perimeter:_____.

6)

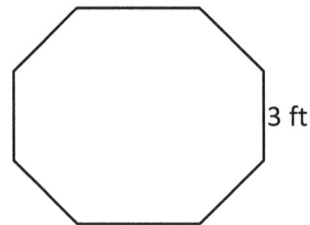

3 ft

Perimeter:_____:

Volume of Cubes

Find the volume of each cube.

1)

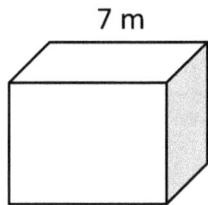

7 m

V:............................

2)

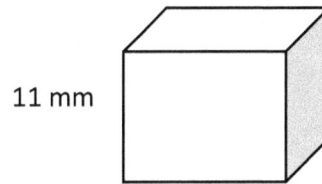

11 mm

V:............................

3)

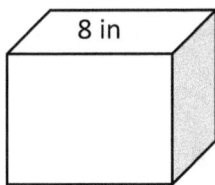

8 in

V:............................

4)

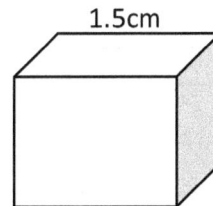

1.5cm

V:............................

5)

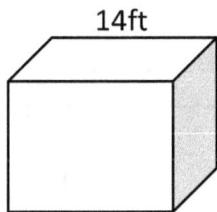

14ft

V:............................

6)

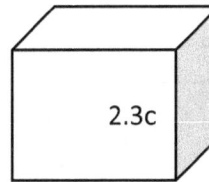

2.3c

V:............................

Volume of Rectangle Prism

Find the volume of each rectangle prism

1)

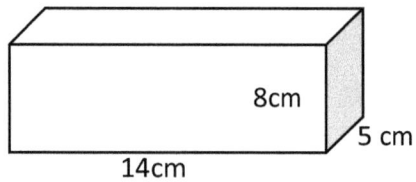

V:..:

2)

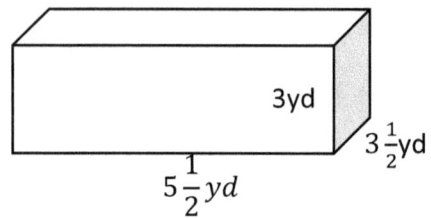

V:..:

3)

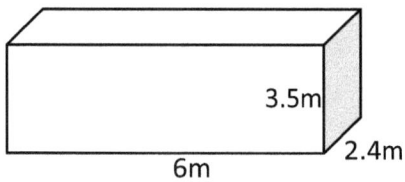

V:..:

4)

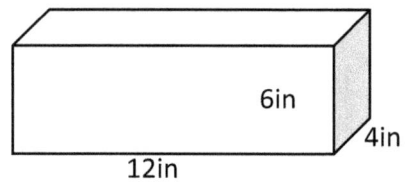

V:..:

5)

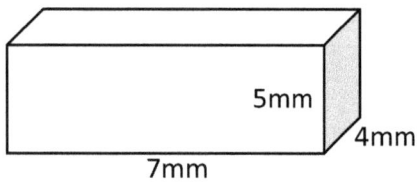

V:..:

6)

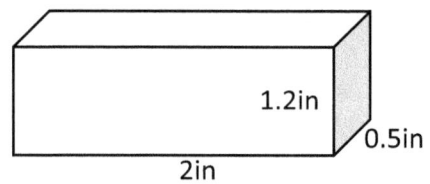

V:..:

Volume of Cylinder

Find the volume of each cylinder.

1)

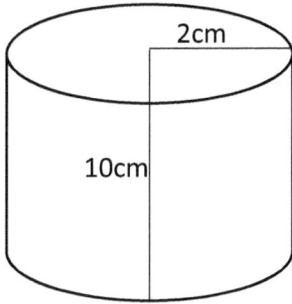

2cm
10cm

V:_____:

2)

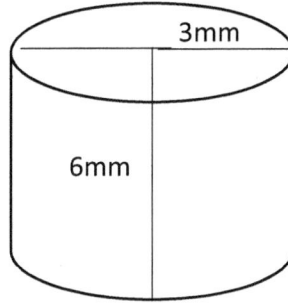

3mm
6mm

V:_____:

3)

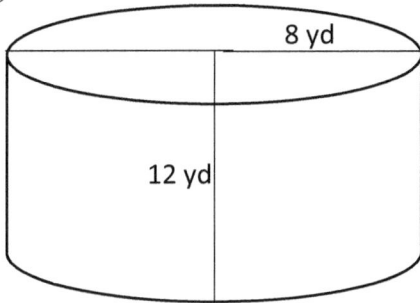

8 yd
12 yd

V:_____:

4)

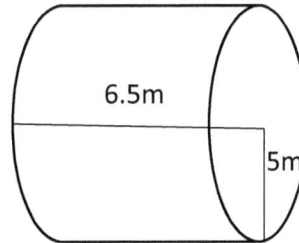

6.5m
5m

V:_____:

5)

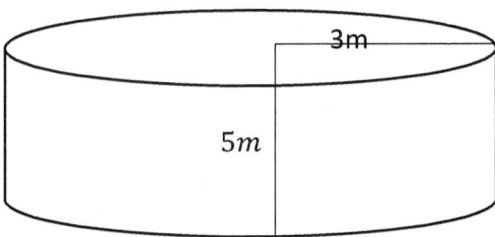

3m
5m

V:_____:

6)

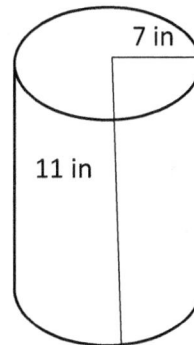

7 in
11 in

V:_____:

Volume of Spheres

Find the volume of each spheres ($\pi = 3.14$).

1)

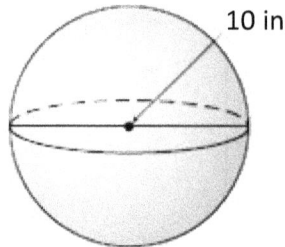

10 in

V:_____:

2)

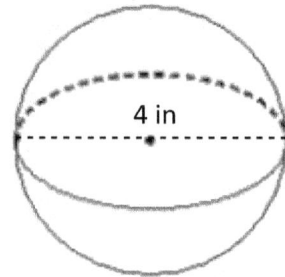

4 in

V:_____:

3)

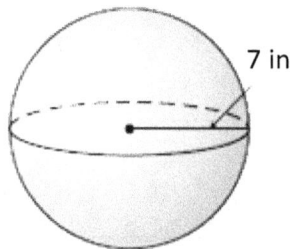

7 in

V:_____:

4)

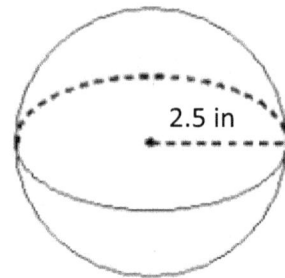

2.5 in

V:_____:

5)

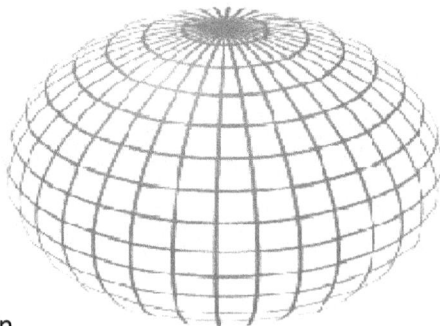

$r = 4\frac{1}{2}$ in

V:_____:

6)

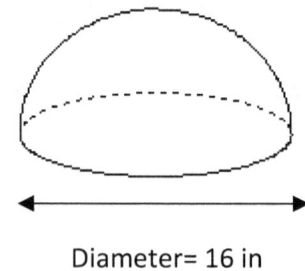

Diameter= 16 in

V:_____:

Volume of Pyramid and Cone

Find the volume of each pyramid and cone ($\pi = 3.14$).

1)

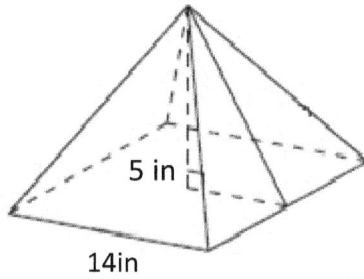

5 in

14in

V:_____.

2)

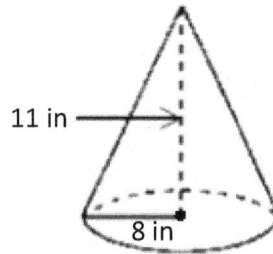

11 in

8 in

V:_____.

3)

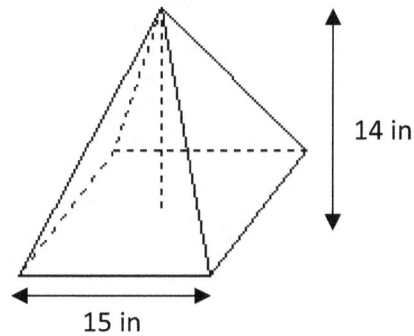

14 in

15 in

V:_____.

4)

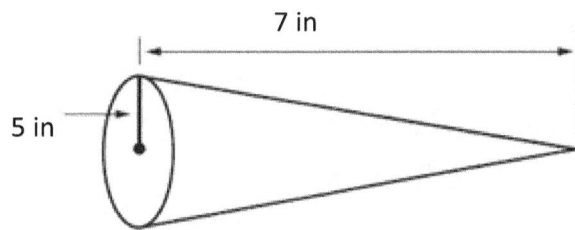

7 in

5 in

V:_____.

5)

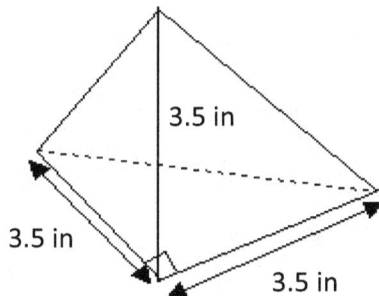

3.5 in

3.5 in

3.5 in

V:_____.

6)

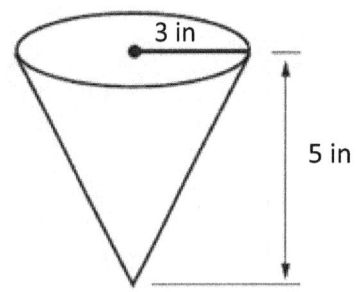

3 in

5 in

V:_____.

Surface Area Cubes

Find the surface area of each cube.

1)

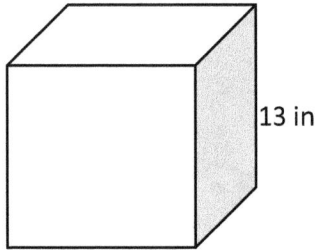

13 in

SA:..:

2)

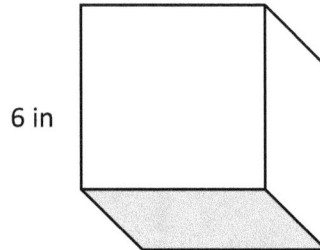

6 in

SA:..:

3)

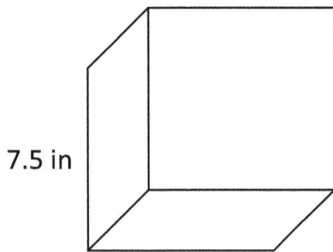

7.5 in

SA:..:

4)

$\sqrt{18}$ in

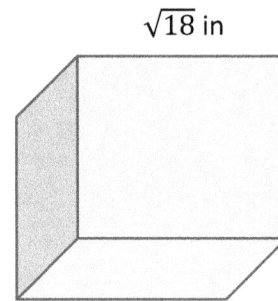

SA:..:

5)

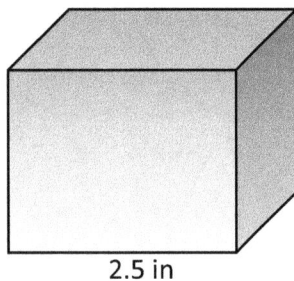

2.5 in

SA:..:

6)

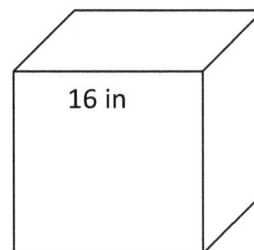

16 in

SA:..:

Surface Area Rectangle Prism

Find the surface area of each rectangular prism.

1)

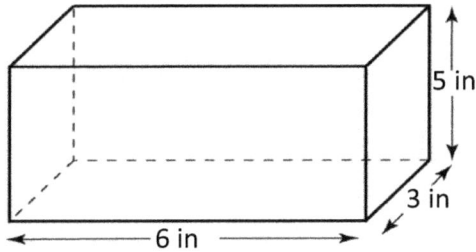

SA:..............................:

2)

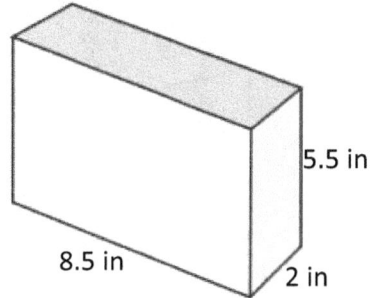

SA:..............................:

3)

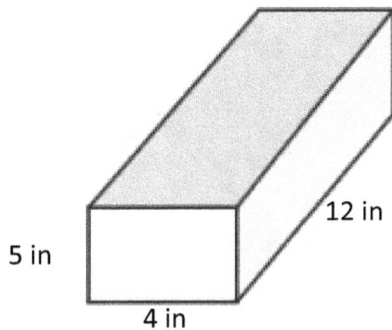

SA:..............................:

4)

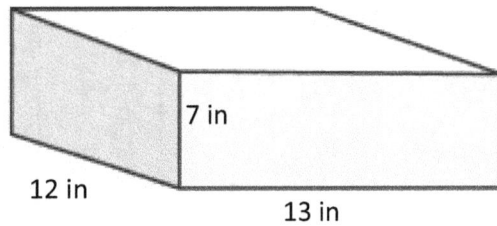

SA:..............................:

5)

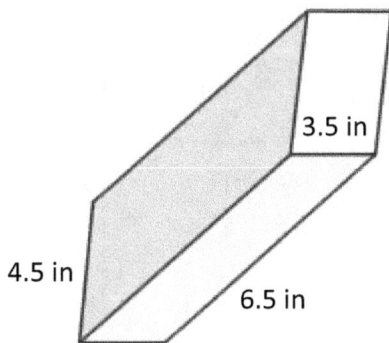

SA:..............................:

6)

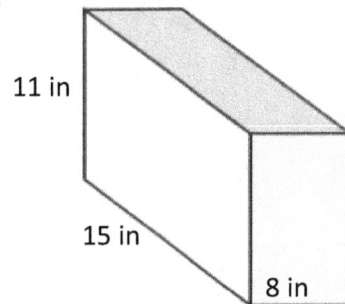

SA:..............................:

Surface Area Cylinder

Find the surface area of each cylinder.

1)

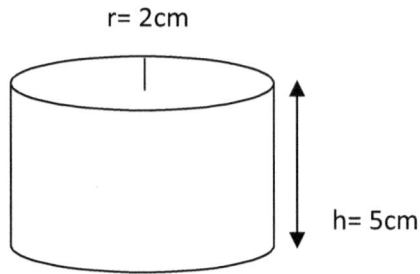

r= 2cm

h= 5cm

SA:_____.

2)

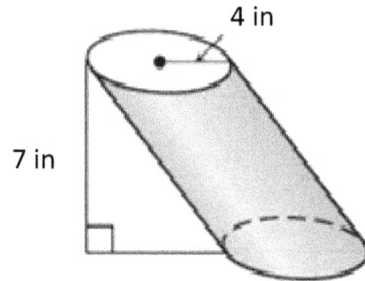

4 in

7 in

SA:_____.

3)

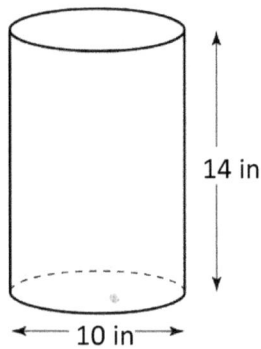

14 in

10 in

SA:_____.

4)

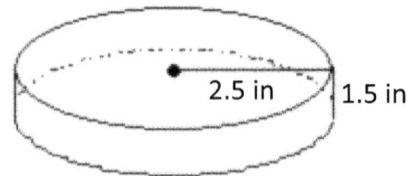

2.5 in   1.5 in

SA:_____.

5)

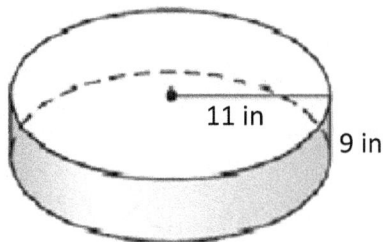

11 in

9 in

SA:_____.

6)

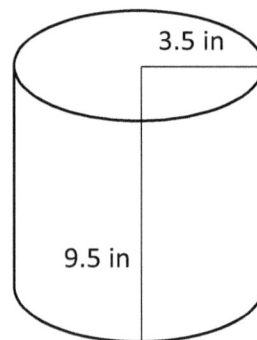

3.5 in

9.5 in

SA:_____.

Answer key Chapter 9

Area and Perimeter of Square

1. Perimeter: 24,   Area:36
2. Perimeter: $4\sqrt{5}$, Area:5
3. Perimeter: 32,   Area:64
4. Perimeter: $4\sqrt{7}$, Area:7
5. Perimeter: 44,   Area:121
6. Perimeter: $4\sqrt{32}$, Area:32

Area and Perimeter of Rectangle

1- Perimeter: 18,   Area:14
2- Perimeter: 30,   Area:50
3- Perimeter: 40,   Area:91
4- Perimeter: 19, Area: 12
5- Perimeter: 12, Area: 8.64
6- Perimeter: 20,   Area:24

Area and Perimeter of Triangle

1- Perimeter: 3s, Area:$\frac{1}{2}sh$
2- Perimeter: 30,   Area:30
3- Perimeter: 32,   Area:32
4- Perimeter: 36,   Area:48
5- Perimeter: 48,   Area:96
6- Perimeter: 60, Area:120

Area and Perimeter of Trapezoid

1- Perimeter: 40,   Area:72
2- Perimeter: 27,   Area:40
3- Perimeter: 41,   Area:62
4- Perimeter: 38,   Area:80
5- Perimeter: 44,   Area:104
6- Perimeter: 60,   Area:192

Area and Perimeter of Parallelogram

1- Perimeter: $26m$, Area:$30(m)^2$
2- Perimeter: $58m$, Area:$120(m)^2$
3- Perimeter: $44in$,   Area:$60(in)^2$
4- Perimeter: $37cm$, Area:$50(cm)^2$
5- Perimeter: $80m$,   Area:$364(m)^2$
6- Perimeter: $60m$, Area:$225(m)^2$

Circumference and Area of Circle

1) Circumference:43.96 mm   Area:$153.86(mm)^2$
2) Circumference: 21.98 in   Area:$38.465(in)^2$
3) Circumference: 13.816 m   Area:$15.197(m)^2$
4) Circumference: 31.4 cm   Area:$78.5(cm)^2$
5) Circumference: 18.84 in   Area:$28.26(in)^2$
6) Circumference: 28.26 km   Area:$63.59(km)^2$

Perimeter of Polygon

1) 55 mm
2) 42 m
3) 65 cm
4) 28 in
5) 40 m
6) 24 ft

Volume of Cubes

1) $343m^3$
2) $1,331(mm)^3$
3) $512in^3$
4) $3.375(cm)^3$

5) $2,744(ft)^3$

6) $12.167(cm)^3$

Volume of Rectangle Prism

1) $560(cm)^3$

2) $57.75(yd)^3$

3) $50.4(m)^3$

4) $288(in)^3$

5) $140(mm)^3$

6) $1.2(in)^3$

Volume of Cylinder

1) $125.6(cm)^3$

2) $42.39(mm)^3$

3) $602.88(yd)^3$

4) $510.25(m)^3$

5) $141.3(m)^3$

6) $1,692.46(in)^3$

Volume of Spheres

1) $523.33(in)^3$

2) $33.49(in)^3$

3) $1,436.02(in)^3$

4) $65.42(in)^3$

5) $381.51(in)^3$

6) $1,071.79(in)^3$

Volume of Pyramid and Cone

1) $326.67\ (in)^3$

2) $736.85\ (in)^3$

3) $1,050\ (in)^3$

4) $183.16\ (in)^3$

5) $7.15\ (in)^3$

6) $47.1(in)^3$

Surface Area Cubes

1) $1,014(in)^2$

2) $216(in)^2$

3) $337.5(in)^2$

4) $108(in)^2$

5) $37.5(in)^2$

6) $1,536(in)^2$

Surface Area Rectangle Prism

1) $126(in)^2$

2) $149.5(in)^2$

3) $256(in)^2$

4) $662(in)^2$

5) $135.5(in)^2$

6) $746(in)^2$

Surface Area Cylinder

1) $87.92(in)^2$

2) $276.32(in)^2$

3) $596.6(in)^2$

4) $62.8(in)^2$

5) $1,381.6(in)^2$

6) $285.74(in)^2$

Chapter 10:

Statistics and probability

Mean, Median, Mode, and Range of the Given Data

Find the mean, median, mode(s), and range of the following data.

1) 24, 59, 20, 37, 14, 24, 47

Mean: __, Median: __, Mode: __, Range: __

2) 6, 13, 13, 19, 15, 10

Mean: __, Median: __, Mode: __, Range: __

3) 21, 35, 49, 11, 45, 27, 35, 19, 14

Mean: __, Median: __, Mode: __, Range: __

4) 25, 11, 1, 15, 25, 18

Mean: __, Median: __, Mode: __, Range: __

5) 24, 14, 14, 17, 23, 15, 14, 29, 29, 8

Mean: __, Median: __, Mode: __, Range: __

6) 7, 14, 19, 11, 8, 19, 8, 15

Mean: __, Median: __, Mode: __, Range: __

7) 29, 28, 66, 76, 14, 44, 18 ,44, 22, 44

Mean: __, Median: __, Mode: __, Range: __

8) 35, 35, 57, 78, 59

Mean: __, Median: __, Mode: __, Range: __

9) 16, 16, 29, 46, 54

Mean: __, Median: __, Mode: __, Range: __

10) 13, 9, 3, 3, 5, 6, 7

Mean: __, Median: __, Mode: __, Range: __

11) 4, 12, 4, 6, 1, 8

Mean: __, Median: __, Mode: __, Range: __

12) 8, 9, 15, 15, 17, 17, 17

Mean: __, Median: __, Mode: __, Range: __

13) 7, 7, 1, 16, 1, 7, 19

Mean: __, Median: __, Mode: __, Range: __

14) 13, 17, 10, 12, 12, 18, 15, 19

Mean: __, Median: __, Mode: __, Range: __

15) 9, 14, 19, 19, 29

Mean: __, Median: __, Mode: __, Range: __

16) 6, 6, 16, 18, 15, 22, 37

Mean: __, Median: __, Mode: __, Range: __

17) 25, 11, 14, 25, 18, 13, 7, 5

Mean: __, Median: __, Mode: __, Range: __

18) 55, 34, 34, 48, 85, 7

Mean: __, Median: __, Mode: __, Range: __

19) 54, 28, 28, 65, 5, 8

Mean: __, Median: __, Mode: __, Range: __

20) 88, 84, 23, 26, 11, 88, 19

Mean: __, Median: __, Mode: __, Range: __

Box and Whisker Plot

1) Draw a box and whisker plot for the data set:

   16, 11, 14, 12, 14, 12, 16, 16, 20

2) The box-and-whisker plot below represents the math test scores of 20 students.

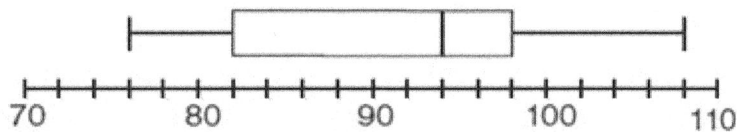

   A. What percentage of the test scores are less than 82?

   B. Which interval contains exactly 50% of the grades?

   C. What is the range of the data?

   D. What do the scores 76, 94, and 108 represent?

   E. What is the value of the lower and the upper quartile?

   F. What is the median score?

Bar Graph

Each student in class selected two games that they would like to play. Graph the given information as a bar graph and answer the questions below:

Game	Votes
Football	13
Volleyball	10
Basketball	18
Baseball	17
Tennis	13

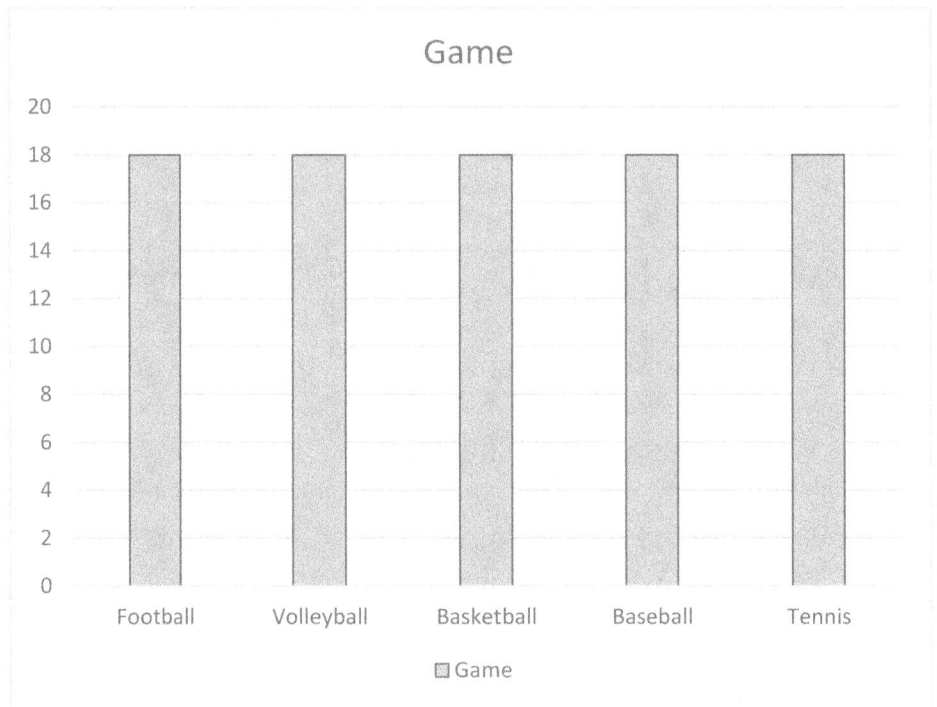

Game

(Bar graph showing all bars at 18 for Football, Volleyball, Basketball, Baseball, Tennis; y-axis from 0 to 20)

☐ Game

1) Which was the most popular game to play?

2) How many more student like Basketball than Volleyball?

3) Which two game got the same number of votes?

4) How many Volleyball and Football did student vote in all?

5) Did more student like football or Volleyball?

6) Which game did the fewest student like?

Histogram

Create a histogram for the set of data.

Math Test Score out of 100 points.

58	74	63	80	83	65	70	86	67	54
81	73	82	75	71	56	87	66	74	72
84	55	76	73	67	85	69	68	52	87

Frequency Table	
Interval	**Number of Values**

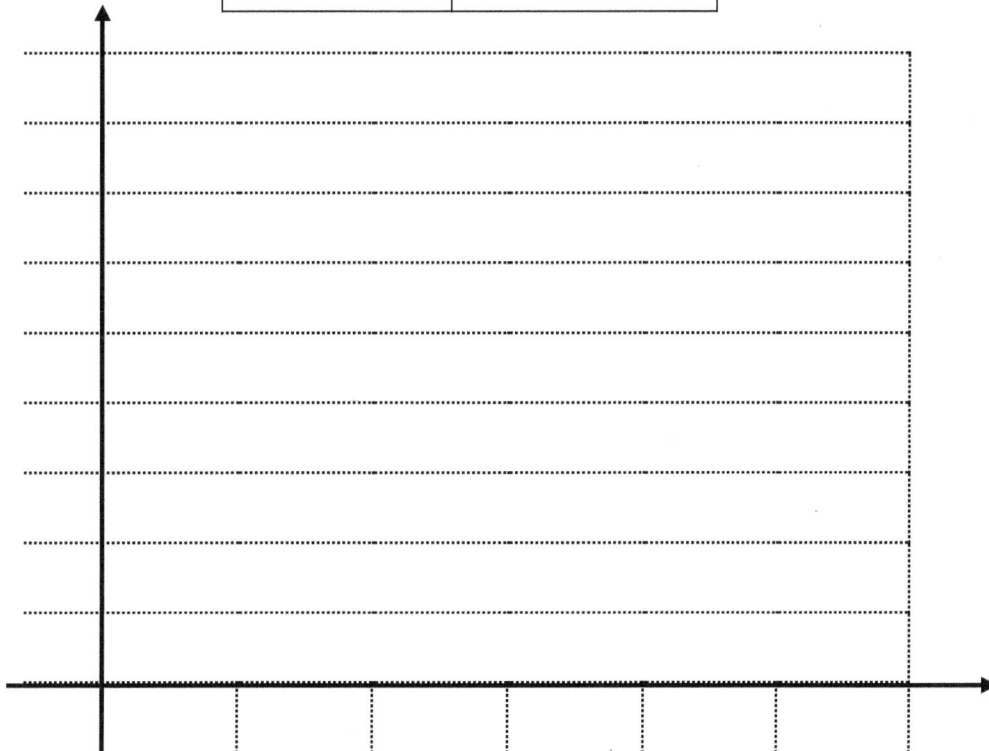

Dot plots

The ages of students in a Math class are given below.

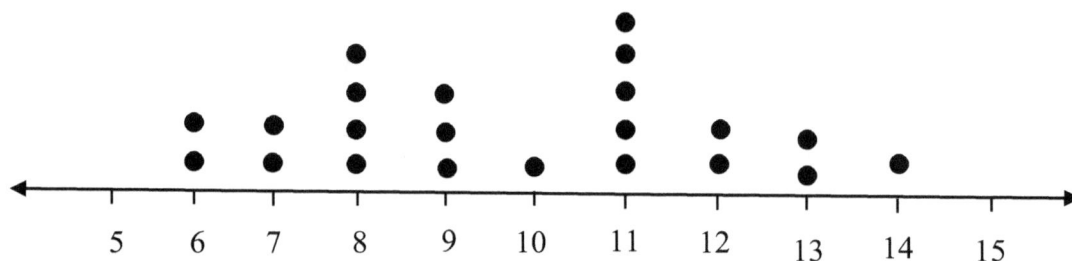

1) What is the total number of students in math class?

2) How many students are at least 12 years old?

3) Which age(s) has the most students?

4) Which age(s) has the fewest student?

5) Determine the median of the data.

6) Determine the range of the data.

7) Determine the mode of the data.

Scatter Plots

A person charges an hourly rate for his services based on the number of hours a job takes.

Hours	Rate
1	$24.5
2	$22
3	$21
4	$19.50

Hours	Rate
5	$19
6	$17.50
7	$17
8	$16.5

1) Draw a scatter plot for this data.

2) Does the data have positive or negative correlation?

3) Sketch the line that best fits the data.

4) Find the slope of the line.

5) Write the equation of the line using slope-intercept form.

6) Using your prediction equation: If a job takes 10 hours, what would be the hourly rate?

Stem–And–Leaf Plot

Make stem-and-leaf plots for the given data.

1) 15, 16, 38, 31, 12, 54, 18, 37, 39, 34, 19, 32, 55

Stem	leaf

2) 72, 74, 17, 41, 72, 14, 46, 78, 48, 44, 49, 42

Stem	leaf

3) 125, 108, 65, 65, 105, 127, 62, 126, 68, 124, 66, 109

Stem	leaf

4) 61, 45, 66, 60, 99, 63, 90, 97, 68, 63, 49, 42

Stem	leaf

5) 55, 58, 105, 56, 15, 108, 102

Stem	leaf

6) 123, 57, 77, 55, 120, 127, 73, 124, 58, 123, 79, 71

Stem	leaf

Pie Graph

60 people were survey on their favorite ice cream. The pie graph is made according to their responses. Answer following questions based on the Pie graph.

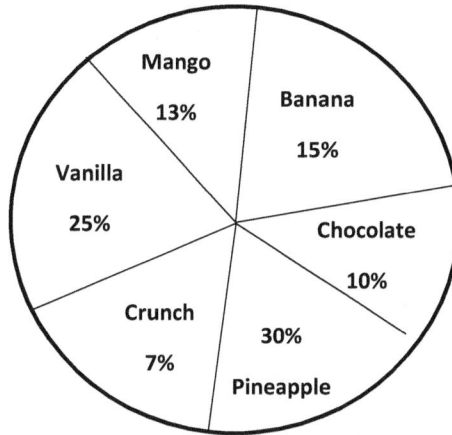

1) How many people like to eat Banana ice cream? _____

2) Approximately, which two ice creams did about half the people like the best? _____

3) How many people said either mango or crunch ice cream was their favorite? _____

4) How many people would like to have chocolate ice cream? _____

5) Which ice cream is the favorite choice of 15 people? _____

Probability

1) A jar contains 16 caramels, 5 mints and 19 dark chocolates. What is the probability of selecting a mint?

2) If you were to roll the dice one time what is the probability it will NOT land on a 4?

3) A die has sides are numbered 1 to 6. If the cube is thrown once, what is the probability of rolling a 5?

4) The sides of number cube have the numbers 4, 6, 8, 4, 6, and 8. If the cube is thrown once, what is the probability of rolling a 6?

5) Your friend asks you to think of a number from ten to twenty. What is the probability that his number will be 15?

6) A person has 8 coins in their pocket. 2 dime, 3 pennies, 2 quarter, and a nickel. If a person randomly picks one coin out of their pocket. What would the probability be that they get a penny?

7) What is the probability of drawing an odd numbered card from a standard deck of shuffled cards (Ace is one)?

8) 32 students apply to go on a school trip. Three students are selected at random. what is the probability of selecting 4 students?

Answer key Chapter 10

Mean, Median, Mode, and Range of the Given Data

1) mean: 32.14, median: 24, mode: 24, range: 45

2) mean: 12.67, median: 13, mode: 13, range: 13

3) mean: 28.44, median: 27, mode: 35, range: 38

4) mean: 15.83, median: 16.5, mode: 25, range: 24

5) mean: 18.7, median: 16, mode: 14, range: 21

6) mean: 12.63, median: 12.5, mode: 19, 8, range: 12

7) mean: 38.5, median: 36.5, mode: 44, range: 62

8) mean: 52.8, median: 57, mode: 35, range: 43

9) mean: 32.2, median: 29, mode: 16, range: 38

10) mean: 6.57, median: 6, mode: 3, range: 10

11) mean: 5.83, median: 5, mode: 4, range: 11

12) mean: 14, median: 15, mode: 17, range: 9

13) mean: 8.29, median: 7, mode: 7, range: 18

14) mean: 14.5, median: 14, mode: 12, range: 9

15) mean: 18, median: 19, mode: 19, range: 20

16) mean: 17.14, median: 16, mode: 6, range: 31

17) mean: 14.75, median: 13.5, mode: 25, range: 20

18) mean: 43.83, median: 41, mode: 34, range: 78

19) mean: 31.33, median: 28, mode: 28, range: 60

20) mean: 48.43, median: 26, mode: 88, range: 77

Box and Whisker Plot

1)

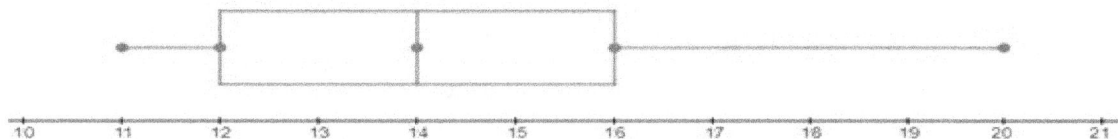

2)

A. 25%

B. 94

C. 32

D. Minimum, Median, and Maximum

E. Lower (Q_1) is 82 and upper (Q_3) is 98

F. 94

Bar Graph

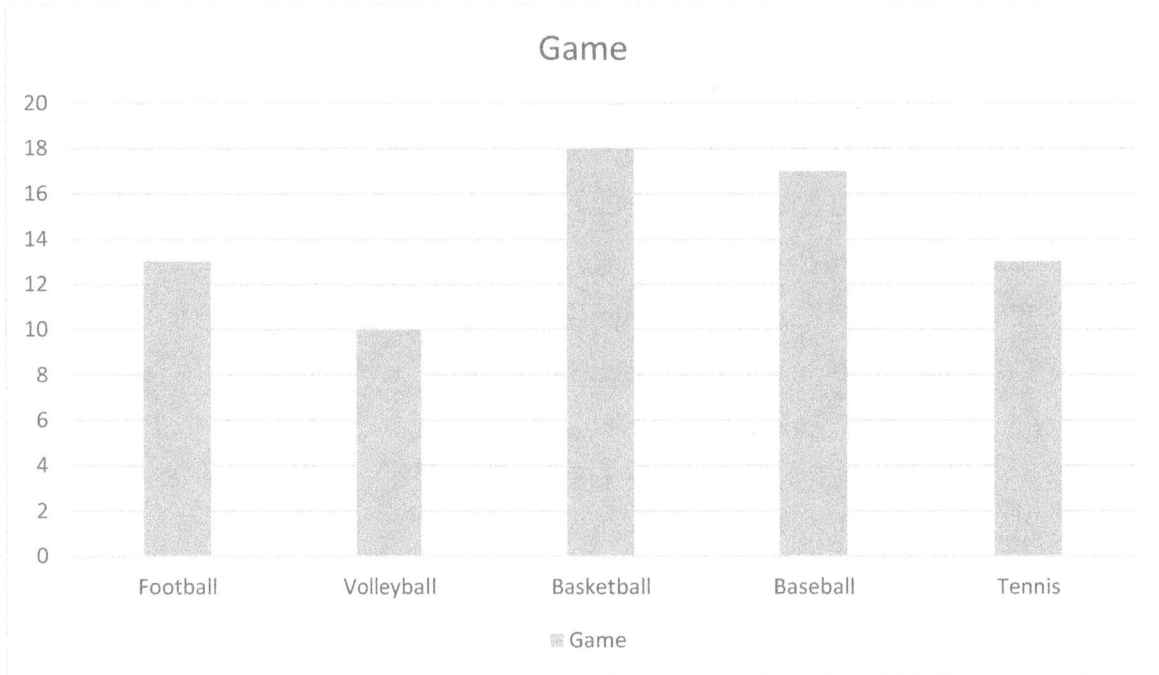

1) Basketball

2) 8 students

3) Football and Tennis

4) 23

5) Football

6) Volleyball

Histogram

Frequency Table	
Interval	**Number of Values**
52-57	4
58-63	2
64-69	6
70-75	8
76-81	3
82-87	7

Histogram

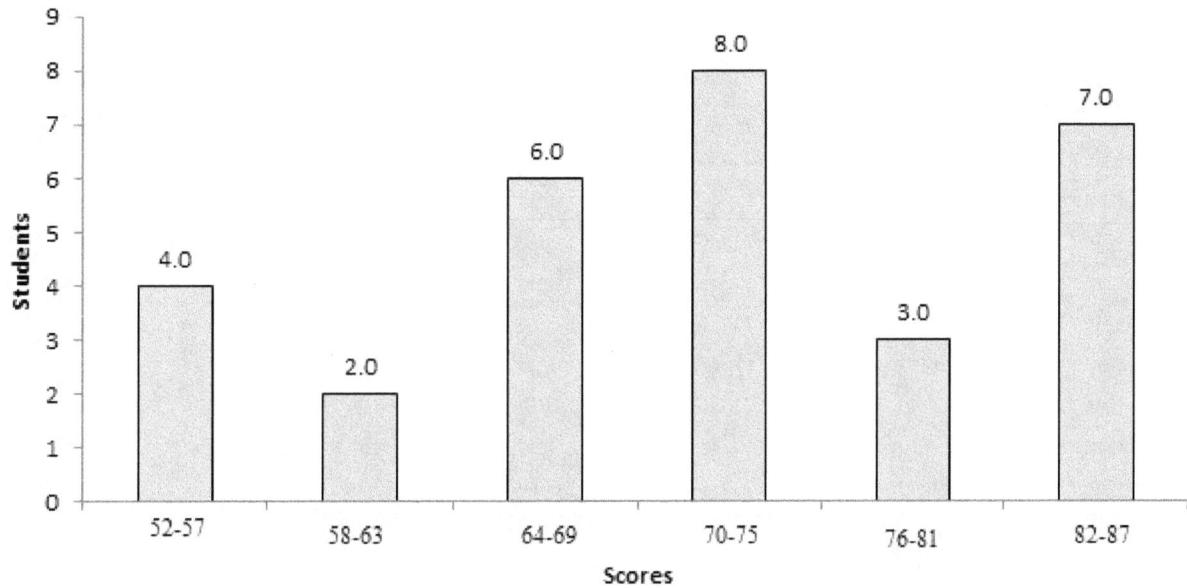

Dot plots

1) 22

2) 5

3) 11

4) 10 and 14

5) 2

6) 4

7) 2

Scatter Plots

1)

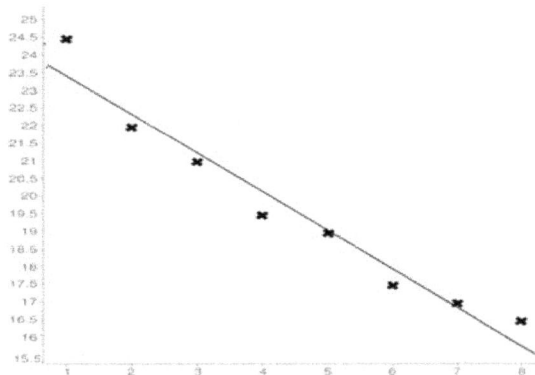

2) Negative correlation

3) ----

4) Slope(m) $= -1$

5) $y = -x + 24.5$

6) 14.5

Stem–And–Leaf Plot

1)

Stem	leaf
1	2 5 6 8 9
3	1 2 4 7 8 9
5	4 5

2)

Stem	leaf
1	4 7
4	1 2 4 6 8 9
7	2 2 4 8

3)

Stem	leaf
6	2 5 5 6 8
10	5 8 9
12	4 5 6 7

4)

Stem	leaf
4	2 9 5
6	0 1 3 3 6 8
9	0 7 9

5)

Stem	leaf
1	5
5	5 6 8
10	2 5 8

6)

Stem	leaf
5	5 7 8
7	1 3 7 9
12	0 3 3 4 7

Pie Graph

1) 9

2) Vanilla and pineapple

3) 12

4) 6

5) Vanilla

Probability

1) $\frac{1}{8}$

2) $\frac{5}{6}$

3) $\frac{1}{6}$

4) $\frac{1}{3}$

5) $\frac{1}{10}$

6) $\frac{3}{8}$

7) $\frac{5}{13}$

8) $\frac{1}{8}$

Mathematics

Test Review

GRADE 8 MAHEMATICS REFRENCE MATERIALS

Linear Equations		
Slope-intercept form		$y = mx + b$
Slope of a line		$m = \frac{y_2 - y_1}{x_2 - x_1}$

Circumference		
Circle	$C = 2\pi r$ or	$C = \pi d$

Area	
Triangle	$A = \frac{1}{2}bh$
Rectangle or Parallelogram	$A = bh$
Trapezoid	$A = \frac{1}{2}h(b_1 + b_2)$
Circle	$A = \pi r^2$

Surface Area	**Lateral**	**Total**
Prism	$S = ph$	$S = ph + 2B$
Cylinder	$S = 2\pi rh$	$S = 2\pi rh + 2\pi r^2$

Volume	
Prism or cylinder	$V = Bh$
Pyramid or Cone	$V = \frac{1}{3}Bh$
Sphere	$V = \frac{4}{3}\pi r^3$

Additional Information	
Pythagorean theorem	$a^2 + b^2 = c^2$
Simple interest	$I = prt$
Compound interest	$I = p(1 + r)^t$

Mathematics Practice Test 1

GRADE 8

Administered *Month Year*

1) Arrange the following fractions in order from least to greatest.

$$\frac{3}{5}, \frac{2}{9}, \frac{1}{8}, \frac{17}{25}, \frac{13}{15}$$

A. $\frac{1}{8}, \frac{2}{9}, \frac{3}{5}, \frac{17}{25}, \frac{13}{15}$

B. $\frac{2}{9}, \frac{3}{5}, \frac{1}{8}, \frac{13}{15}, \frac{17}{25}$

C. $\frac{17}{25}, \frac{13}{15}, \frac{2}{9}, \frac{3}{5}, \frac{1}{8}$

D. $\frac{13}{15}, \frac{17}{25}, \frac{1}{8}, \frac{3}{5}, \frac{2}{9}$

2) Elena earns $8.50 an hour and worked 28 hours. Her brother earns $11.90 an hour. How many hours would her brother need to work to equal Elena's earnings over 28 hours?

A. 16.22

B. 20

C. 12

D. 10.50

3) In a library, 30% of the books are fiction and the rest are non-fiction. Given that there are 1,600 more non-fiction books than fiction books, what is the total number of books in the library?

A. 2,600

B. 3,800

C. 4,000

D. 1,200

4) Which of the following expressions is undefined in the set of real numbers?

    A. $\sqrt[2]{156}$

    B. $\sqrt[3]{-64}$

    C. $\sqrt{-49}$

    D. $\sqrt[5]{32}$

5) If $f(x) = 2x^2$, and $2f(3a) = 144$ then what could be the value of a?

    A. -2

    B. -4

    C. 4

    D. 2

6) What is the value of 2^6?

    A. $(3 + 3)^2$

    B. 4^3

    C. $2(2^3)$

    D. $2^3 + 2^3$

7) Shane Williams puts $4,500 into a saving bank account that pays simple interest of 5.2%. How much interest will she earn after 4 years?

    A. $5,480

    B. $ 1,580

    C. $936

    D. $148

8) A map has the scale of 6 cm to 1 km. What is the actual area of a lake on ground which is represented as an area of 162 cm^2 on the map?

A. 2.7 km^2

B. 4.5 km^2

C. 45 cm^2

D. 27 cm^2

9) Which equation can be equal "6 more than the ratio of a number to 8 is equal to 9 less than the number"?

A. $6x - 8 = 9 - x$

B. $6 + \frac{x}{8} = x - 9$

C. $\frac{6}{8}x - 9 = 8x$

D. $6 + 8x = 9 - x$

10) Three angles join to form a straight angle. One angle measure 68°. Other angle measures 37°. What is the measure of third angle?

A. 15°

B. 31°

C. 65°

D. 75°

11) Evaluate $\dfrac{35x^6y^5z^{-3}}{10\,x^3y^7z^2}$.

   A. $\dfrac{5x^3y^2}{2\,z^5}$

   B. $\dfrac{5x^3z^5}{2\,y^2}$

   C. $\dfrac{7x^3}{2\,y^2z^5}$

   D. $\dfrac{7y^2}{2\,x^3z^5}$

12) Find the equation for line passing through $(1, -2)$ and $(3, 3)$.

   A. $-5x - 2y = 19$

   B. $2y - 5x = -9$

   C. $-5x + 2y = 19$

   D. $2y + 5x = -9$

13) Which of the following equations best represents the line in the graph below?

   A. $y = \dfrac{1}{2}\,x + 3$

   B. $y = x + 7$

   C. $y = \dfrac{1}{2}\,x - 3$

   D. $y = x + 3$

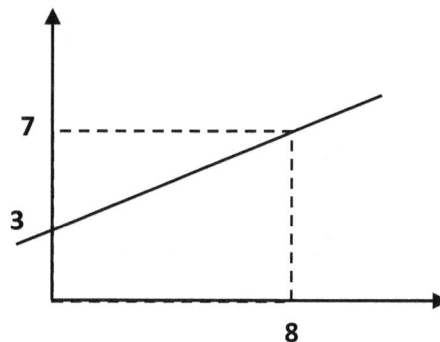

14) lengths of Two sides of a triangle are 4 and 7. Which of the following could

Not be the measure of third side?

A. 2

B. 5

C. 8

D. 10

15) Rosie is x years old. She is 2 years older than her twin brothers Milan and

Marcel. What is the mean age of the two children?

A. $x + 1$

B. $x - 4$

C. $x + 4$

D. $x - 1$

16) a is inversely proportional to $(3b - 8)$. If a = 14 and b = 6, express a in terms

of b.

A. $a = 3b - 8$

B. $a = 145(3b - 8)$

C. $a = \frac{140}{3b-8}$

D. $a = \frac{3b-8}{140}$

17) The table shows the parking rates for the outside terminal area at an airport.

Ethan parked at the lot for $5\frac{1}{2}$ hours. How much did he owe?

A. $5

B. $4.50

C. $7

D. $7.50

3 hours	$4.5
Each 30 minutes after 3 hours	$0.50
24-hours Discount rate	$38

18) The circle graph below shows the type of pizza that people prefer for lunch. If

260 people were surveyed, how many people preferred Meat lover's?

A. 15

B. 24

C. 39

D. 78

Pizza

Other 13%

Margherita 25%

Meat-Lover's 15%

Veggie 17%

Peperoni 30%

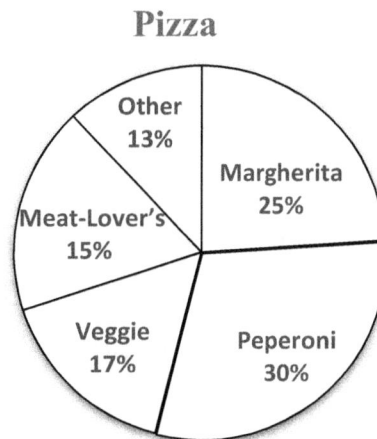

19) A baseball has a volume of 288π. What is the length of the diameter?

A. 6

B. 12

C. 18

D. 24

20) If $-5x + 2y = -7$ and $6x - 3y = 6$, what is the value of x?

    A. -3

    B. 2

    C. -2

    D. 3

21) Solve the linear inequality: $-\frac{(6x-10)}{5} + 7 \geq 9$

    A. $x < 0$

    B. $x \leq 0$

    C. $x > 0$

    D. $x \geq 0$

22) The line n has a slope of $\frac{c}{d}$, where c and d are integers. What is the slope of a line that is perpendicular to line n?

    A. $-\frac{d}{c}$

    B. $\frac{c}{d}$

    C. $\frac{d}{c}$

    D. $-\frac{c}{d}$

23) Which graph represents a linear relationship?

A.

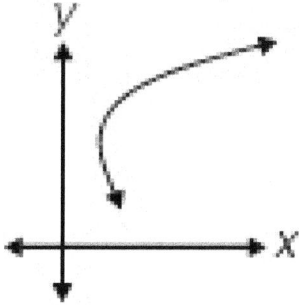

B.

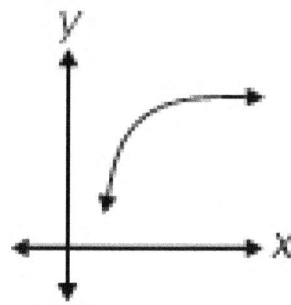

C.

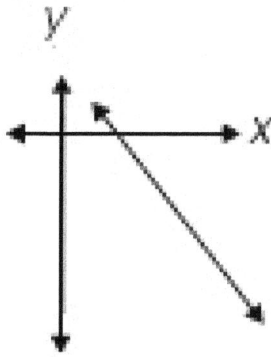

D.

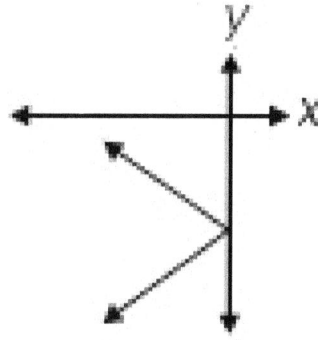

24) In the infinitely repeating decimal below, 8 is the second digit in the repeating pattern. What is the 512st digit? $\frac{2}{7} = \overline{0.285714}$

A. 1

B. 2

C. 14

D. 8

25) Express as a single fraction in its simplest form: $\dfrac{4}{(x-1)} - \dfrac{7}{(2x+3)} = ?$

    A. $\dfrac{x+19}{(x-1)(2x+3)}$

    B. $\dfrac{3x-19}{(x-1)(2x+3)}$

    C. $\dfrac{x+19}{2x-3}$

    D. $\dfrac{-19}{x-2}$

26) Ryan is x years old and her sister Mitzi is $(7x - 16)$ years old. Given that

Mitzi is triple as old as Ryan, what is Mitzi's age?

    A. 8

    B. 4

    C. 12

    D. 10

27) Which table of values represents a linear function?

A)

x	y
-1	1
0	3
1	6
2	8

B)

x	y
-1	1
0	2
1	5
2	7

C)

x	y
-1	1
0	3
1	5
2	8

D)

x	y
-1	1
0	3
1	5
2	7

28) Right triangle XYZ and points J and K are shown below.

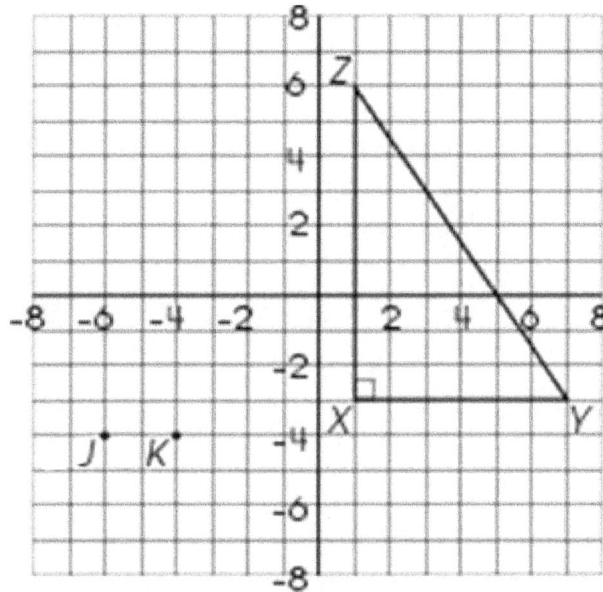

Triangle XYZ is similar to triangle JKL. What could be the location of point L so that triangle JKL is similar to triangle XYZ?

A. $(-4, -6)$

B. $(6, -1)$

C. $(-6, -1)$

D. $(-4, 1)$

29) For what value(s) of x is the following equation true: $2x^2 - 12x + 18 = 0$?

A. $2, 3$

B. -3

C. $+3$

D. ± 3

30) The scatter plot in the figure below shows the GPA of the students of a class versus their commute time. Which of the following statements is true?

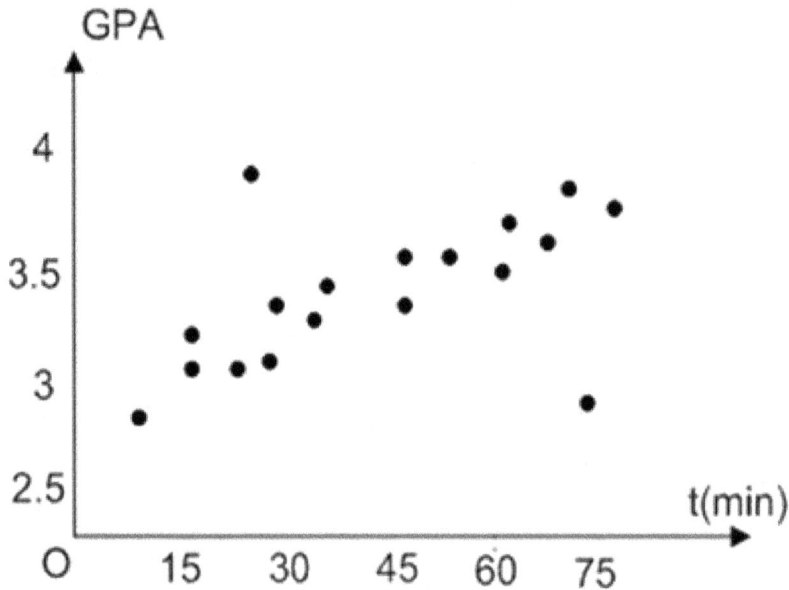

A. The relationship between GPA and commute time is moderate positive linear.

B. The relationship between GPA and commute time is moderate negative linear.

C. There is no relation between GPA and commute time.

D. The scatter plot does not have any outliers.

31) If $x = -2$ and $y = 2$, calculate the value of $\frac{x^2+5}{y+1}$.26

A. $\frac{1}{3}$

B. -3

C. 3

D. 2

32) $12.24 \div 0.4 = 13$

    A. 3.06

    B. 30.60

    C. 30.06

    D. 3.006

33) Alfred needs to calculate his monthly water bill. His family used 23,700

    gallons at a rate of $0.95 per hundred gallons. Also, there is a monthly fee of

    $4.20 on each period. What is his total bill? 15

    A. $22,519.20

    B. $225.15

    C. $229.35

    D. $6.45

34) What is the value of x in term of c and d $\frac{c-d}{dx} = \frac{2}{3}$? (c and d>0) 28

    A. $\frac{3}{2}\left(\frac{c}{d} - 1\right)$

    B. $\frac{3}{2}\left(1 - \frac{c}{d}\right)$

    C. $\frac{2}{3}\left(\frac{c}{d} - 1\right)$

    D. $\frac{2}{3}\left(1 - \frac{c}{d}\right)$

35) The following data set is given: 134, 118, 148, 184, 159, 151.

Adding which number to the set will increase its mean? 34

A. 129

B. 139

C. 149

D. 159

36) In a store, 38% of customers are female. If the total number of customers is

950, then how many male customers are dealing with the store? 53

A. 228

B. 361

C. 589

D. 675

37) Evaluate: $\frac{(x^2+5x+6)}{(2x^2-8x+8)} \div \frac{(x^2+2x-3)}{(x^2-3x+2)} =$? 52

A. $\frac{x+2}{2x-2}$

B. $\frac{2(x+2)}{x-2}$

C. $\frac{x-2}{2x-4}$

D. $\frac{x+2}{x-2}$

38) 221 is What percent of 170? 43

  A. 130 %

  B. 77 %

  C. 30 %

  D. 123 %

39) The set of possible values of p is {2,5,11}. What is the set of possible values

  of h if $3h = 2p + 2$? 57

  A. {2,5,11}

  B. {6,15,33}

  C. {2,4,8}

  D. {6,12,24}

40) Solve for y: $1.87 - 0.6y = -0.86$. 62

  A. 4.25

  B. 4.65

  C. 4.55

  D. 3.75

"End of Practice Test 1."

Mathematics Practice Test 2

GRADE 8

Administered *Month Year*

1) What is the sum of the smallest prime number and five times the largest negative even integer?

   A. -4

   B. 3

   C. -8

   D. -6

2) What is the number a, if the result of adding a to 36 is the same as subtracting $5a$ from 300?

   A. 84

   B. 264

   C. 144

   D. 44

3) Solve these fractions and reduce to its simplest terms: $2\frac{7}{20} - 3\frac{3}{5} + 1\frac{1}{2} =$

   A. $\frac{1}{4}$

   B. $-1\frac{1}{5}$

   C. $\frac{4}{5}$

   D. $-1\frac{3}{5}$

4) Find the solution set of the following equation: $|4x - 9| = 11$

    A. $\{5, -0.5\}$

    B. $\{4, -0.2\}$

    C. $\{0.5\}$

    D. $\{-5, 0.5\}$

5) What is the solution to the pair of equations below? $\begin{cases} 3x + 2y = 5 \\ 4x - y = 3 \end{cases}$

    A. $x = 1$ and $y = -1$

    B. $x = -1$ and $y = 1$

    C. $x = 1$ and $y = 1$

    D. $x = 1$ and $y = 0$

6) What is $\sqrt[5]{5^{-10}}$ in simplest form?

    A. $\dfrac{1}{3,125}$

    B. $\dfrac{1}{625}$

    C. $\dfrac{1}{125}$

    D. $\dfrac{1}{25}$

7) What is 9.36×10^{-4} in standard form?

    A. $-93,600$

    B. -0.000936

    C. $\dfrac{1}{93,600}$

    D. 0.000936

8) What is the value of x in the triangle?

A. 141°

B. 103°

C. 39°

D. 119°

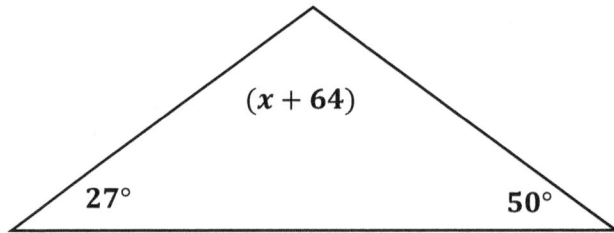

9) Find the length of the unknown side.

A. 22.6 ft

B. 10 ft

C. 100 ft

D. 35.40 ft

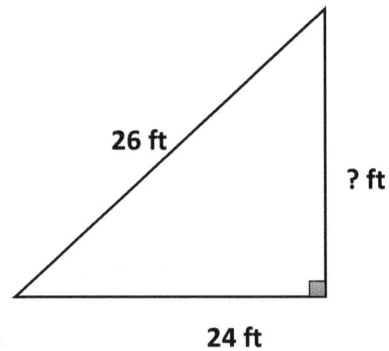

10) A store sells all of its products at a price 25% greater than the price the store paid for the product. How much does the store sell a product if the store paid $140 for it?

A. $105

B. $210

C. $175

D. $35

11) Which statement correctly describes the value of N in the equation below?

$$5(6N - 15) = 6(5N - 15)$$

A. N=0 is one solution.

B. N has infinitely many correct solutions.

C. N has no correct solutions.

D. N=1 is one solution.

12) What is the probability of Not spinning at H?

A. $\frac{3}{8}$

B. $\frac{5}{8}$

C. $\frac{2}{3}$

D. $\frac{1}{8}$

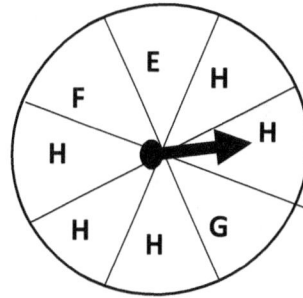

13) A position of subway station and Grace's house shown by a grid. The station is located at $(-3, -5)$, and her house is located at $(1, -2)$. What is the distance between her house and the subway stop?

A. 7

B. 5

C. $5\sqrt{2}$

D. 15

14) Each of 7 pitchers can contain up to $\frac{5}{7}$ L of water. If each of the pitcher is at least the half full, which of the following expressions represents the total amount of water, W, contained on all 7 pitchers?

A. $0.5 < w < 5$

B. $0 < w < 2.5$

C. $0 < w < 5$

D. $2.5 < w < 5$

15) What is the simplest form of the expression $\frac{(4x^{-3}y^2z)^2}{80y^{-3}z^3}$, (using positive exponent)?

A. $\frac{2y^7}{5x^4z^2}$

B. $\frac{y^7}{5x^6z}$

C. $\frac{x^5z}{5y^6}$

D. $\frac{x^5}{5y^7z}$

16) What is the simplest form of the expression $\frac{3x^2-14x-5}{9(x^2-\frac{1}{9})}$?

A. $\frac{x+5}{3(x-\frac{1}{3})}$

B. $\frac{x-5}{3x-1}$

C. $\frac{x+5}{3x+1}$

D. $\frac{3x-1}{x-5}$

17) The graph below shows the diameters of a number of red oak trees versus their ages.

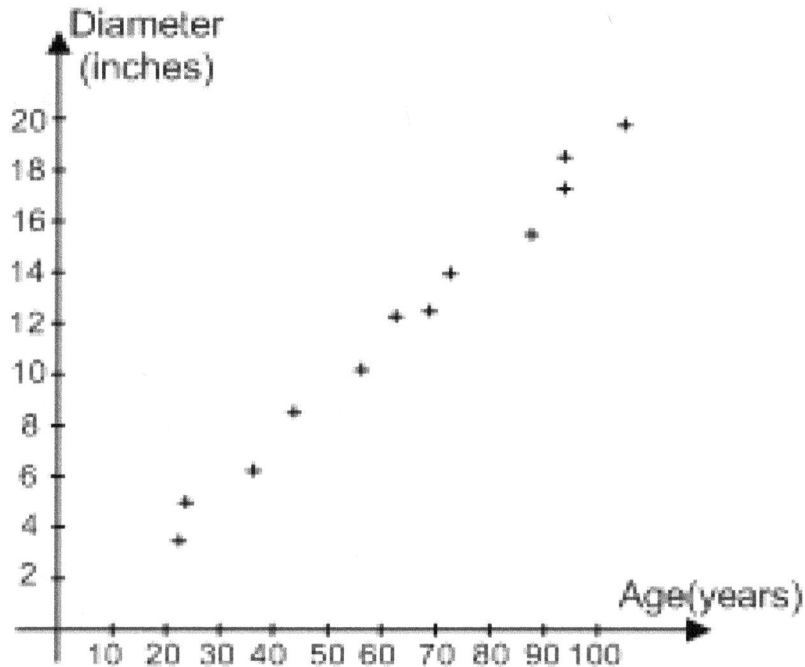

Assuming a linear growth rate, what is the diameter of a 160-year-old red oak tree?

A. 32 inches

B. 30 inches

C. 16 inches

D. 28 inches

18) The price of a shirt increased from $40 to $41.60. What is the percentage increase in the price?

A. 4%

B. 0.4%

C. 0.94%

D. 1.4%

19) The line $3y - 2 = 9x + 13$ and $3y - 3 = x + 1$ are?

    A. Perpendicular

    B. Parallel

    C. The same line

    D. Neither parallel nor perpendicular

20) Find the slope of the line.

    A. $\dfrac{1}{3}$

    B. $\dfrac{1}{6}$

    C. 3

    D. $-\dfrac{1}{3}$

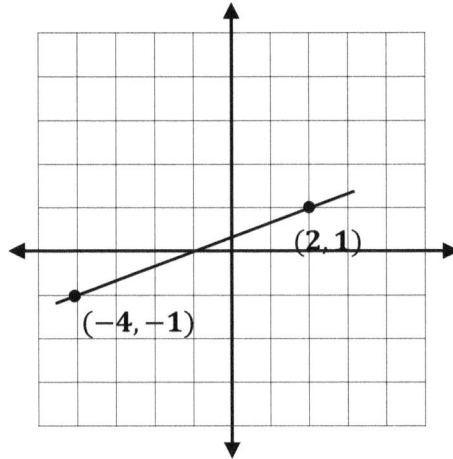

$(2, 1)$

$(-4, -1)$

21) Let $f(x) = 4x - 3$. If $f(a) = -11$ and $f(b) = 17$, then what is $f(a + b)$?

    A. -6

    B. 6

    C. 9

    D. -9

22) if $xy - 8x = 45$ and $y - 8 = 9$, then $x =$?

    A. 45

    B. 15

    C. 9

    D. 5

23) What is the number of sides of a regular polygon whose interior angles

are 156° each? (Remember, the sum of exterior angles of any polygon

is 360°).

A. 10

B. 15

C. 8

D. 24

24) The sum of two consecutive integers is −13. If 1 is added to the smaller

integer and 3 is subtracted from the larger integer, what is the product of the

two resulting integers?

A. 54

B. 42

C. 48

D. 63

25) $6 + (7n + 5) - (8n + 3)$

A. $12 - n$

B. $12 + n$

C. $8 + n$

D. $8 - n$

26) If the line m is parallel to the side BC of ABC, what is angle n?

A. 115

B. 57

C. 25

D. 15

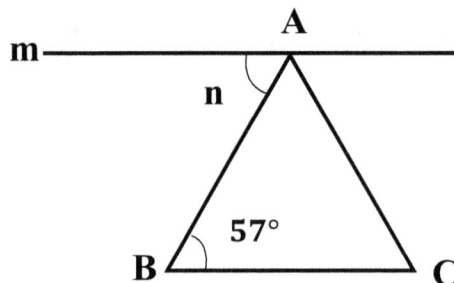

27) The operation is defined as $a \,@\, b = a - ab$

The operation # is defined as $a\#b = 2a^2 - b$

If $f(x) = -2x^2 - 2$, what is the value of $f(1)\#\big(f(1)\,@\,f(-2)\big)$?

A. -4

B. -12

C. -68

D. 36

28) What is the maximum amount of grain, the silo can hold, in cubic feet?

A. $60\pi \; m^3$

B. $324\pi \; m^3$

C. $384\pi \; m^3$

D. $1{,}024\pi \; m^3$

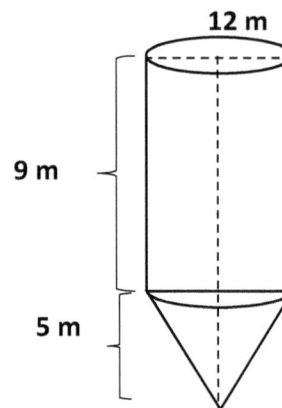

29) Triangle MRQ is shown on the coordinate grid. A student reflected Triangle MRQ across the y −axis to create Triangle M'R'Q'.

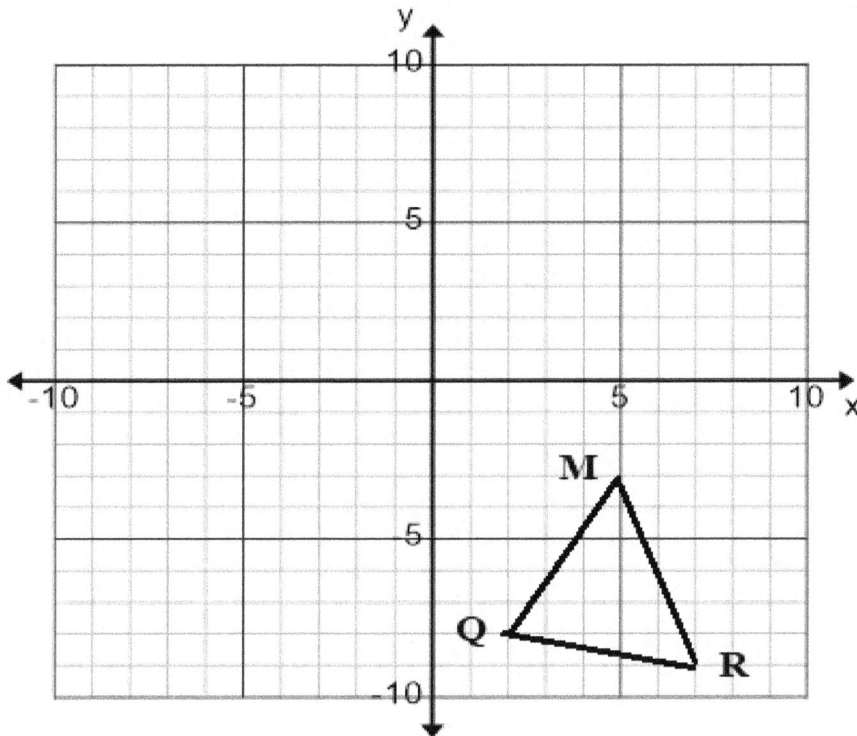

Which rule describes this transformation?

A. $(x, y) \rightarrow (-x, y)$

B. $(x, y) \rightarrow (x, y + 5)$

C. $(x, y) \rightarrow (y, -x)$

D. $(x, y) \rightarrow (x, -y)$

30) The equation $x = 2y - 4$ has a y-intercept of? 31

A. $\frac{1}{2}$

B. 2

C. $-\frac{1}{4}$

D. -4

31) The figure shows line MN parallel to line EF. The lines are intersected by 2 transversals. All lines are in the same plane.

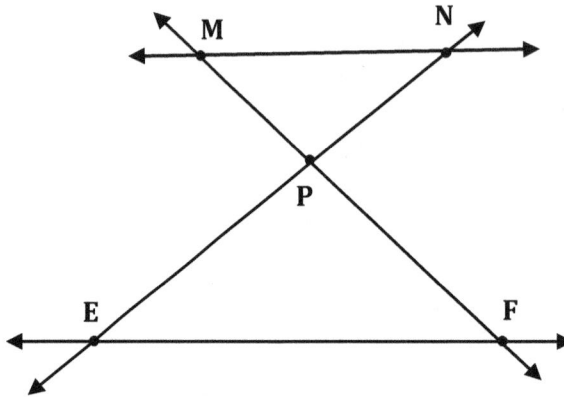

If $m\angle$MPE = 115, Determine $m\angle$PNM + $m\angle$PFE?

A. 115

B. 100

C. 65

D. 25

32) The price of water doubles every 5 years. If the price of water on January 1st, 2012 is \$2 per gallon, what is the equation that would be used to calculate the price(P) of water on January 1st, 2007? 45

A. $5P = 2$

B. $\dfrac{P}{2} = 2$

C. $2p = 2$

D. $2P = 5$

33) What is solution to the equation $\sqrt{3x - 1} = 4$? 30

    A. −1

    B. −5

    C. 1

    D. 5

34) In a competition, three teams, FT, RT and GT, scored a total of 160 points.

    If FT scored 45% of this total and RT scored three times as many points

    as GT, what were the number of points scored by RT? 53

    A. 66

    B. 72

    C. 6

    D. 22

35) Amelia cuts a piece of birthday cake as shown below. What is the volume of

    the piece of cake? 47

    A. 540 cm^3

    B. 270 cm^3

    C. 180 cm^3

    D. 2,430 cm^3

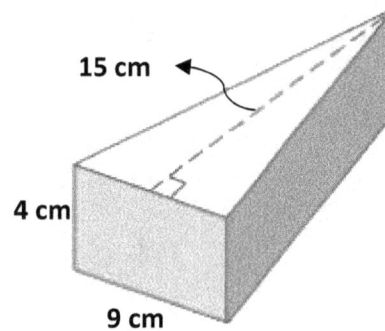

15 cm

4 cm

9 cm

36) If $3^{2x} = 81$, then $x = ?$ 32

    A. 4

    B. 3

    C. 2

    D. 1

37) What is the percent equivalent of 0.004? 61

    A. 40%

    B. 4%

    C. 0.40%

    D. 0.04%

38) When a number is subtracted from 90 and the result is divided by 7, the

    answer obtained is twice the original number. What is the number? 62

    A. 6

    B. 9

    C. 10

    D. 7

39) Some fruit sells for $10 per kilograms. What is the price in cent per gram? 42

    A. 0.001

    B. 0.1

    C. 0.01

    D. 1

40) A sports store has a container of handballs: 7 blue, 10 red, 11 yellow, 8 white, and 4 green. If one ball is picked from the container at random, what is the probability that it will be green? 55

A. $\frac{1}{4}$

B. $\frac{1}{10}$

C. $\frac{1}{9}$

D. $\frac{9}{10}$

"End of Practice Test 2."

Answers and

Explanations

Answer Key

Now, it's time to review your results to see where you went wrong and what areas you need to improve!

Math Practice Tests

Practice Test 1

1	A	16	C	31	C
2	B	17	C	32	B
3	C	18	C	33	C
4	C	19	B	34	A
5	D	20	D	35	D
6	B	21	B	36	C
7	C	22	A	37	C
8	B	23	C	38	A
9	B	24	A	39	C
10	D	25	A	40	C
11	C	26	C		
12	B	27	D		
13	A	28	C		
14	A	29	C		
15	D	30	A		

Practice Test 2

1	C	16	B	31	A
2	D	17	A	32	C
3	A	18	A	33	D
4	A	19	D	34	A
5	C	20	A	35	B
6	D	21	C	36	C
7	D	22	D	37	C
8	C	23	B	38	A
9	B	24	A	39	D
10	C	25	D	40	B
11	C	26	B		
12	A	27	B		
13	B	28	C		
14	D	29	A		
15	B	30	B		

Practice Test 1

Answers and Explanations

1) Answer: A

Rewriting each fraction with common denominator or converting each fraction to decimal and order the decimal from least to greatest.

$\frac{3}{5} = 0.6$ $\qquad$ $\frac{2}{9} = 0.22$ $\qquad$ $\frac{1}{8} = 0.125$ $\qquad$ $\frac{17}{25} = 0.68$ $\qquad$ $\frac{13}{15} = 0.86$

2) Answer: B

calculating Elena's total earnings: 28 hours × \$8.50 an hour = \$238

Next, divide this total by her brother's hourly rate: \$238 ÷ \$11.90 = 20 hours

3) Answer: C

number of fiction books: x

number of nonfiction books: $x + 1,600$

Total number of books: $x + (x + 1,600)$

30% of the total number of books are fiction, Therefore:

$30\%[x + (x + 1,600)] = x \rightarrow 0.3(2x + 1,600) = x$

$0.6x + 480 = x \rightarrow 480 = x - 0.6x \rightarrow 0.4x = 480 \rightarrow x = 1,200$ number of fictions

$x + 1,600 = 1,200 + 1,600 = 2,800$, the number of nonfiction books

$1,200 + 2,800 = 4,000$, the total number of books in the library

4) Answer: C

For odd index we can have negative radicand.

In the even index, negative radicand is undefined.

$\sqrt{-49}$ has a negative number under the even index, so it is non-real.

Negative numbers don't have real square roots, because negative and positive integer squared is either positive or 0.

5) Answer: D

$2f(3a) = 144 \rightarrow$ (divide by 2): $f(3a) = 72$ (subtitute 3a) $2(3a)^2 = 72$

$\rightarrow$ (divide by 2): $(3a)^2 = 36 \rightarrow 9a^2 = 36 \rightarrow a^2 = 4 \rightarrow a = 2$

6) Answer: B

Use formula to raise a number: $(x^a)^b = x^{ab}$

$2^6 = (2^2)^3 = 4^3$

7) Answer: C

Simple interest rate: I = prt (I = interest, p = principal, r = rate, t = time)

$I = 4,500 \times 0.052 \times 4 = 936$

8) Answer: B

The scale is: 6 cm:1km, (6 cm on the map represents an actual distance of 1 km).

first necessary to rewrite the scale ratio in terms of units squared:

6^2cm square:1^2km square, which gives: $36\ cm^2 : 1\ km^2$

Then, $\dfrac{36\ cm^2}{1 km^2} = \dfrac{162\ cm^2}{x\ km^2}$, (where x is the unknown actual area).

Every proportion you write should maintain consistency in the ratios described

(km^2 both occupy the denominator).

Cross-multiply and isolate to solve for the unknown area x:

$x.\dfrac{36\ cm^2}{1 km^2} = 162\ cm^2 \rightarrow x = 162\ cm^2.\dfrac{1\ km^2}{36\ cm^2} \rightarrow x = 4.5\ km^2$

9) Answer: B

6 more: +6

Ratio: ÷ ; Ratio of a number to 8: $\dfrac{x}{8}$

9 less: -9 ; 9 less than the number: $x - 9$

$6 + \dfrac{x}{8} = x - 9$

10) Answer: D

A straight angle is an angle measured exactly 180°

$$68° + 37° = 105°$$

$$180° - 105° = 75°$$

11) Answer: C

$\dfrac{35x^6y^5z^{-3}}{10\ x^3y^7z^2} = \dfrac{35}{10} \times \dfrac{x^6}{x^3} \times \dfrac{y^5}{y^7} \times \dfrac{z^{-3}}{z^2} = \dfrac{7}{2} \times x^3 \times \dfrac{1}{y^2} \times \dfrac{1}{z^5} = \dfrac{7x^3}{2\ y^2z^5}$

12) Answer: B

$$m = \frac{y_2 - y_1}{x_2 - x_1} = \frac{3 - (-2)}{3 - 1} = \frac{5}{2}$$

$$y - y_1 = m(x - x_1) \rightarrow y - (-2) = \frac{5}{2}(x - 1)$$

$$y + 2 = \frac{5}{2}(x - 1) \rightarrow 2(y + 2) = 5(x - 1) \rightarrow 2y + 4 = 5x - 5$$

$$2y - 5x = -5 - 4 \rightarrow 2y - 5x = -9$$

13) Answer: A

Two points are $(0, 3)$ and $(8, 7)$: $m = \frac{y_2 - y_1}{x_2 - x_1} = \frac{7 - 3}{8 - 0} = \frac{4}{8} = \frac{1}{2}$

$$y - y_1 = m(x - x_1) \rightarrow y - 3 = \frac{1}{2}(x - 0)$$

$$y - 3 = \frac{1}{2}x \rightarrow y = \frac{1}{2}x + 3$$

14) Answer: A

Triangle third side rule: length of the one side of a triangle is less than the sum of the lengths of the other two sides and greater than the positive difference of the lengths of the other two sides.

the third side is less than $4 + 7 = 11$ and greater than $7 - 4 = 3$

15) Answer: D

Using the formula for mean:

$$\text{Mean} = \frac{sum\ of\ the\ several\ given\ values}{number\ of\ value\ given} = \frac{x + (x - 2)}{2} = \frac{2x - 2}{2} = \frac{2(x - 1)}{2} = x - 1$$

16) Answer: C

a value that is inversely proportional to another value:

$$a = \frac{k}{3b - 8} \text{ (where k is a constant of proportionality)}$$

substitute a and b: $14 = \frac{k}{3(6) - 8} \rightarrow k = 14(10) = 140$

$$a = \frac{140}{3b - 8}$$

17) Answer: C

Ethan paid $4.5 for three hours and $0.50 for each of five half-hour period after that.

$$5 \times 0.5 = 2.50; \quad 4.5 + 2.50 = 7$$

18) Answer: C

Change percent to decimal: $15\% = 0.15$

$0.15 \times 260 = 39$

19) Answer: B

Baseballs and basketballs are spherical. The volume of sphere: $v = \frac{4}{3}\pi r^3$

$288\pi = \frac{4}{3}\pi r^3 \to 288 = \frac{4}{3}r^3 \to 864 = 4r^3 \to r^3 = 216 \to r = 6$

$d = 2r \to d = 2 \times 6 = 12$

20) Answer: D

$\begin{cases} 3 \times (-5x + 2y = -7) \\ 2 \times (6x - 3y = 6) \end{cases} \to \begin{cases} -15x + 6y = -21 \\ 12x - 6y = 12 \end{cases}$ →add two equations:

$-3x = -9 \to x = 3$

21) Answer: B

subtracting 7 from both sides: $-\frac{(6x-10)}{5} \geq 2$

Multiply both sides by -5: $(6x - 10) \leq -10$

This operation clears the negative sign and the denominator of 5 from the left side. Add 10 to both sides and divide by 6 to isolate x: $6x \leq 0 \to x \leq 0$

22) Answer: A

A line perpendicular to a line with slope m has a slope of $-\frac{1}{m}$.

So, the slope of the line perpendicular to the given line is $-\frac{1}{\frac{c}{d}} = -\frac{d}{c}$.

23) Answer: C

A constant rate of change without a change in direction (a single straight line). Answer choice C is a single straight line and represents a linear relationship. Answer choices A and B are both curves and not linear. Answer choice D contains a constant rate of change; however, the graph is not a single straight line.

24) Answer: A

There are 6 digits in the repeating decimal (285714), so 8 would be the second, eighth, fourteenth digit and so on. To find the 512st digit, divide 512 by 9.

$512 \div 9 = 56$ R8. Since the remainder is 8, that means the 512st digit is the same as the 1st digit, which is 8.

25) Answer: A

Since this is subtraction of 2 fractions with different denominators, their least common denominator is: $(x - 1)(2x + 3)$

$$\frac{4}{(x-1)} - \frac{7}{(2x+3)} = \frac{4(2x+3)-7(x-1)}{(x-1)(2x+3)} = \frac{8x+12-7x+7}{(x-1)(2x+3)} = \frac{x+19}{(x-1)(2x+3)}$$

26) Answer: C

State the problem in a mathematical equation:

$$3x = 7x - 16 \rightarrow 3x - 7x = -16 \rightarrow -4x = -16 \rightarrow x = \frac{-16}{-4} = 4$$

$$7x - 16 = 7(4) - 16 = 12$$

27) Answer: D

A linear function is the one where the slope, a difference of any two y – values over the difference of their corresponding x – values remain the same.

Using data from the table to calculate the slope as the difference of any two y – values they pick from the table over the difference of their corresponding x – values. Answer choice D is correct:

The slope $= \frac{3-1}{0-(-1)} = \frac{2}{1} = 2$; the slope $= \frac{5-3}{1-0} = \frac{2}{1} = 2$,

or the slope $= \frac{7-5}{2-1} = \frac{2}{1} = 2$. The slope remains the same.

28) Answer: C

First determine that the ratio between the original triangle and the similar triangle is 3:1. Using this ratio the length of side LJ would be 3 units. One of the points three units away from point J $(-6, -4)$ would be $(-6, -1)$, Answer choice B.

29) Answer: C.

For a value of x to satisfy the provided equation, it must be a solution. The equation provided should be recognized as a quadratic equation, which can be factored using many methods.

2 can be divided out of both sides: $2x^2 - 12x + 18 = 0$,

$$x^2 - 6x + 9 = 0$$

In order to further simplify this expression and solve for x, we must factor. We are looking for 2 numbers that, when multiplied together, yield +9, and when added together yield −6.

+9 factors: $\pm(1 \times 9), \pm(3 \times 3)$

$(x - 3)(x - 3) = 0 \rightarrow x - 3 = 0 \rightarrow x = 3$

So, x must equal +3. This answer can be confirmed by substituting +3 into the original equation: $3^2 - 6(3) + 9 = 9 - 18 + 9 = 0$

30) Answer: A

The scatter plot has outliers, so c is not the correct answer. Between the other two choices, the relationship between GPA and commute time is moderate positive linear is the correct answer.

31) Answer: C

$$\frac{x^2 + 5}{y + 1} = \frac{(-2)^2 + 5}{2 + 1} = \frac{9}{3} = 3$$

32) Answer: B

move decimal point in divisor so last digit is in the unit place (0.4 to 4)

move decimal point in dividend same number of places to the right,

(12.24 to 122.4), then $122.4 \div 4 =$

divide $(1224 \div 4) = 306$

insert a decimal point into the answer above the decimal point in the dividend (30.6)

33) Answer: C

Be careful with the conversion factor (per hundred gallons; NOT per gallon).

$$23{,}700 \times \frac{0.95}{100} = 225.15$$

$$225.15 + 4.20 = 229.35$$

34) Answer: A

Cross multiply and isolate x: $\frac{c - d}{dx} = \frac{2}{3} \rightarrow 2dx = 3(c - d) \rightarrow x = \frac{3(c - d)}{2d}$

$$x = \frac{3}{2}\left(\frac{c}{d} - \frac{d}{d}\right) = \frac{3}{2}\left(\frac{c}{d} - 1\right)$$

35) Answer: D

Mean = (118 + 134 + 148 + 151 + 159 + 184) ÷ 6 = 894 ÷ 6 = 149

Only 159 can increase the mean.

36) Answer: C

If 38% of the total number of customers is female, then 100% − 38% = 62% of the customers are male. Calculate 62% of the total.

0.62 × 950 = 589.

37) Answer: C.

Recall that dividing fractions is the same as multiplying the first fraction by its reciprocal (numerator and denominator are switched).

First rewrite the expression as a multiplication problem, switching numerator with denominator in the second fraction. Then, factor and simplify, where possible:

$$\frac{(x^2+5x+6)}{(2x^2-8x+8)} \times \frac{(x^2-3x+2)}{(x^2+2x-3)} = \frac{(x+2)(x+3)}{2(x-2)^2} \times \frac{(x-2)(x-1)}{(x-1)(x+3)} = \frac{x+2}{2(x-2)} = \frac{x+2}{2x-4}$$

38) Answer: A

Use percent formula: $\text{Part} = \frac{\text{percent} \times \text{whole}}{100}$

$221 = \frac{\text{percent} \times 170}{100} \Rightarrow \frac{221}{1} = \frac{\text{percent} \times 170}{100}$, cross multiply.

22,100 = percent × 170, divide both sides by 170. → percent = 130

39) Answer: C

$3h = 2p + 2$:

$p = 2 \to 3h = 2(2) + 2 = 6 \to 3h = 6 \to h = 2$

$p = 5 \to 3h = 2(5) + 2 = 12 \to 3h = 12 \to h = 4$

$p = 11 \to 3h = 2(11) + 2 = 24 \to 3h = 24 \to h = 8$

possible values of h: {2,4,8}

40) Answer: C

To solve for y, subtracting 3.98 from both sides of the equation:

$-0.8y = -0.86 - 1.87 \to -0.6y = -2.73$

Divide both sides by -0.6: $y = 4.55$

Practice Test 2

Answers and Explanations

1) Answer: C

The smallest prime number is 2, and the largest even negative integer is −2.

$2 + 5(-2) = 2 - 10 = -8$.

2) Answer: D

State the problem in a mathematical sentence:

$a + 36 = 300 - 5a$

$a + 5a = 300 - 36$

$6a = 264 \rightarrow a = 44$

3) Answer: A

$2\frac{7}{20} - 3\frac{3}{5} + 1\frac{1}{2} = (2 - 3 + 1)\frac{7}{20} - \frac{12}{20} + \frac{10}{20} = (0)\frac{17}{20} - \frac{12}{20} = \frac{5}{20} = \frac{1}{4}$

4) Answer: A

$|4x - 9| = 11 \rightarrow \begin{cases} 4x - 9 = 11 \rightarrow 4x = 20 \rightarrow x = 5 \\ 4x - 9 = -11 \rightarrow 4x = -2 \rightarrow x = -0.5 \end{cases}$

5) Answer: C

Multiply equation (2) by 2. Add two equations [(1) +2(2)]:

$\begin{cases} 3x + 2y = 5 \\ 8x - 2y = 6 \end{cases} \rightarrow 11x = 11 \rightarrow x = 1$

Substitute $x = 1$ into equation (1): $3(1) + 2y = 5 \rightarrow 2y = 5 - 3 = 2 \rightarrow y = 1$

6) Answer: D

$\sqrt[5]{5^{-10}} = \sqrt[5]{\frac{1}{5^{10}}} = \frac{\sqrt[5]{1}}{\sqrt[5]{5^{10}}} = \frac{1}{5^{\left(\frac{10}{5}\right)}} = \frac{1}{5^2} = 5^{-2} = \frac{1}{25}$

7) Answer: D

$9.36 \times 10^{-4} = 0.000936$

8) Answer: C

$x + 64 + 27 + 50 = 180 \rightarrow x + 141 = 180 \rightarrow x = 39$

9) Answer: B

use the Pythagorean theorem to find the value of unknown side.

$a^2 + b^2 = c^2 \rightarrow 26^2 = a^2 + 24^2 \rightarrow a^2 = 676 - 576 = 100 \rightarrow a = 10$

10) Answer: C

Use percent formula: $\text{Part} = \frac{\text{percent} \times \text{whole}}{100}$

$\text{Part} = \frac{25 \times 140}{100} = 35$

Last price: $140 + 35 = \$175$

11) Answer: C

There are no values of the variable that make the equation true.

12) Answer: A

There are 5 parts labeled "H" out of a total of 8 equal parts.

The probability of not spinning at "H" is 3 out of 8.

13) Answer: B

Point $1(x_A, y_A) = (-3, -5)$

Point $2(x_B, y_B) = (1, -2)$

Distance between two points $= \sqrt{(x_B - x_A)^2 + (y_B - y_A)^2}$

$\rightarrow d = \sqrt{\left(1 - (-3)\right)^2 + \left(-2 - (-5)\right)^2} = \sqrt{4^2 + 3^2} = \sqrt{16 + 9} \rightarrow d = \sqrt{25} = 5$

14) Answer: D

The minimum amount of water: $7 \times \frac{5}{7} = 5; \ 5 \div 2 = 2.5$

The maximum amount of water: $7 \times \frac{5}{7} = 5$

Amount of water in all pitchers: $2.5 < w < 5$

15) Answer: B

$$\frac{(4x^{-3}y^2z)^2}{80y^{-3}z^{-3}} = \frac{16x^{-6}y^4z^2}{80y^{-3}z^3} = \frac{y^7}{5x^6z}$$

16) Answer: B

Factor the expression: $\frac{3x^2 - 14x - 5}{9(x^2 - \frac{1}{9})} = \frac{(x-5)(3x+1)}{9x^2 - 1} = \frac{(x-5)(3x+1)}{(3x-1)(3x+1)} = \frac{x-5}{3x-1}$

17) Answer: A

The line has an intercept equal to zero and a slope of $\frac{20 \ inches}{100 \ years} = 0.2$

The growth factor of red oak trees is $\frac{1}{0.2} = 5$ years/inch.

The diameter of a 160-year-old red oak tree is $\frac{160}{5} = 32$ inches.

18) Answer: A

Use the formula for Percent of Change:

$\frac{New\ Value - Old\ Value}{Old\ Value} \times 100\ \% = \frac{41.60 - 40}{40} \times 100\% = \frac{1.60}{40} \times 100\% = 4\%$

19) Answer: D

First equation: $3y - 2 = 9x + 13 \rightarrow 3y = 9x + 15 \rightarrow y = 3x + 5 \rightarrow m_1 = 3$

Second question: $3y - 3 = x + 1 \rightarrow 3y = x + 4 \rightarrow y = \frac{1}{3}x + \frac{4}{3} \rightarrow m_2 = \frac{1}{3}$

$m_1 = 3$ and $m_2 = \frac{1}{3}$, they aren't equal slopes or negative reciprocals.

20) Answer: A

Two points are $(2, 1)$ and $(-4, -1) \rightarrow m = \frac{y_2 - y_1}{x_2 - x_1} = \frac{-1 - 1}{-4 - 2} = \frac{-2}{-6} = \frac{1}{3}$

21) Answer: C

f(a) = 4a − 3 and f(a) = −11: 4a − 3 = −11 so that 4a = −8 and a = −2.

f(b) = 4b − 3 and f(b) = 17: 4b − 3 = 17 so that 4b = 20 and b = 5

Finally, f(a + b) = f(−2 + 5) = f(3), f(3) = 4(3) − 3 = 9.

22) Answer: D

$xy - 8x = 45 \rightarrow x(y - 8) = 45$; substitute $y - 8 = 9 \rightarrow 9x = 45 \rightarrow x = 5$

23) Answer: B

The formula for measurement of each angle of a regular polygon:

$x = \frac{180(n-2)}{n}$, (x is the measurement of the interior angle; n is the number of sides)

Substituting the given information:

$\frac{180(n - 2)}{n} = 156 \rightarrow 180n - 360 = 156n \rightarrow 180n - 156\,n = 360$

$\rightarrow 24\,n = 360 \rightarrow n = 15$. The polygon has 15 angles and 15 sides.

24) Answer: A

If x is the smaller consecutive integer, then $x + 1$ is the larger consecutive integer. Use their sum (-13) to find x:

$$x + (x + 1) = -13 \rightarrow 2x + 1 = -13 \rightarrow 2x = -14 \rightarrow x = -7$$

The two consecutive integers are -7 and -6.

One is added to the smaller integer: $-7 + 1 = -6$

and 3 is subtracted from the larger integer: $-6 - 3 = -9$

Find the product: $(-6)(-9) = 54$

25) Answer: D

$$6 + (7n + 5) - (8n + 3) = 6 + 7n + 5 - 8n - 3 = 8 - n$$

26) Answer: B

Two parallel lines (m & side BC) intersected by side AB. $n = 57°$ (interior angles)

27) Answer: B

Note: "@" and "#" are two "operations", and they are defined in the first two lines of the question.

Begin by evaluating $f(1)$ and $f(-2)$, Then, substitute $f(1)$ and $f(-2)$ into the first operation and substitute the appropriate values into the second operation.

$$f(1) = -2(1)^2 - 2 = -2 - 2 = -4$$

$$f(-2) = -2(-2)^2 - 2 = -8 - 2 = -10$$

$a @ b = a - ab$:

$$(f(1) @ f(-2)) = (-4) @ (-10) = (-4) - (-4)(-10) = -4 - 40 = -44$$

$a\#b = 2a^2 - b$:

$$f(1)\#(f(1) @ f(-2)) = (-4)\#(-44) = 2(-4)^2 - 44 = 32 - 44 = -12$$

28) Answer: C

Volume of cylinder: $V = \pi r^2 h = \pi \times 6^2 \times 9 = 324\pi$

Volume of cone: $V = \frac{1}{3}\pi r^2 h = \frac{1}{3}\pi \times 6^2 \times 5 = 60\pi$

$324\pi + 60\pi = 384\pi$.

29) Answer: A

when a point is reflected (flipped) across the $x-$axis (horizontal), the $y-$coordinate (vertical position from 0) is multiplied by -1. Therefore, the rule $(x, y) \to (-x, y)$ describes the transformation (to change a shape using a rotation (a circular movement), reflection (flip), translation (slide), or dilation (resize)).

30) Answer: B

Get the equation into slop intercept form:

$y = mx + b$ where m is slope and b is y-intercept.

$x = 2y - 4$; Add 4 to both sides $\to x + 4 = 2y$; Dividing by 2: $\to y = \frac{1}{2}x + 2$

Thus, the y-intercept is 2.

31) Answer: A

Angles PMN and PFE are alternate interior angles so m∠ PMN= m∠PFE.

Since m ∠ MPN+∠ m MPE $= 180$ and m ∠ MPE $= 115°$, m ∠ MPN$= 180° - 115° =$ 65°. The measures of the angles of a triangle sum to 180° so, m∠ PNM $+$ m ∠ PMN $=$ 180° $-$ m∠ MPN $= 180°- 65° = 115°$ So m ∠PNM $+$ m∠ PFE $=$ m∠ PNM $+$ m∠ PMN= 115°.

32) Answer: C

Area$=9\frac{1}{2} \times 7\frac{1}{5} = \frac{19}{2} \times \frac{36}{5} = \frac{684}{10} = \frac{342}{5}$ square feet.

One-fourth $= \frac{342}{5} \div 4 = \frac{342}{5} \times \frac{1}{4} = \frac{342}{20} = \frac{171}{10} = 17.10$

33) Answer: D

$\sqrt{3x - 1} = 4 \to 3x + 1 = 16 \to 3x = 15 \to x = 5$

34) Answer: A

Let f be the points earned by FT, r be the points scored by RT, and g be the points scored by GT.

f $= 45\%$ of $160 = 0.45 \times 160 = 72$ points

$160 = f + r + g \to 160 = 72 + 3g + g \to 160 - 72 = 4g$

$\rightarrow 4g = 88 \rightarrow g = 22$

$r = 3g \rightarrow r = 3(22) = 66$ points

35) Answer: B

$V = \left(\frac{base \times height}{2}\right) \times$ height of prism $\rightarrow V = \left(\frac{9 \times 15}{2}\right) \times 4 = 270 \; cm^3$

36) Answer: C

$3^{2x} = 81 \rightarrow (3^2)^x = 81 \rightarrow 9^x = 9^2 \rightarrow x = 2$

37) Answer: C

To write a percent, move the decimal point two places to the right and follow the resulting number with the % sign. Or adding the % sign after multiplying the decimal number and 100: $0.004 \times 100 = 0.4\% = 0.40\%$

38) Answer: A

$\frac{(90-x)}{7} = 2x \rightarrow 90 - x = 7(2x) \rightarrow 90 = 14x + x \rightarrow x = \frac{90}{15} = 6$

39) Answer: D

Rate: $\frac{\$1}{1Kg} = \frac{100¢}{1,000g} \rightarrow 1\;^{\$}/_{kg} = 0.1\;^{¢}/_{g}$

The price is: $10 \times 0.1 = 1$ cent per kilogram.

40) Answer: B

The total number of handballs in the container is.

$$7 + 10 + 4 + 8 + 11 = 40$$

Since there are 4 yellow handballs, the probability of selecting a green handball is

$\frac{4}{40} = \frac{1}{10}$

"END"